Fifty Ships

that Changed the Course of

History

A FIREFLY BOOK

Published by Firefly Books Ltd. 2016

Copyright © 2016 Quid Publishing

First printing

Publisher Cataloging-in-Publication Data (U.S.)

Names: Graham, Ian, 1953– , author.
Title: Fifty ships that changed the course of history : a nautical history of the world / Ian Graham.
Description: Richmond Hill, Ontario, Canada : Firefly Books, 2016. | Includes bibliography and index. | Summary: "A visual history of economic development in fifty ships, starting from the earliest known record, Pharaoh Khufu's solar barge (roughly 5000 years ago), to MS Allure of the Seas, the biggest passenger ship ever built (longer than the Eiffel Tower is tall)" — Provided by publisher.
Identifiers: ISBN 978-1-77085-719-3 (hardcover)
Subjects: LCSH: Shipping — History. | Shipping – Economic aspects.
Classification: LCC HE571.G734 |DDC 387.5 – dc23

Library and Archives Canada Cataloguing in Publication

A CIP record for this title is available from Library and Archives Canada

Published in the United States by
Firefly Books (U.S.) Inc.
P.O. Box 1338, Ellicott Station
Buffalo, New York 14205

Published in Canada by
Firefly Books Ltd.
50 Staples Avenue, Unit 1
Richmond Hill, Ontario L4B 0A7

Interior design: Lindsey Johns

Printed in China

Conceived, designed and
produced by
Quid Publishing
Level 1, Ovest House
West Street
1 2RA
nd
www.quidpublish

Fifty Ships

that Changed the Course of

History

A NAUTICAL HISTORY OF THE WORLD

Ian Graham

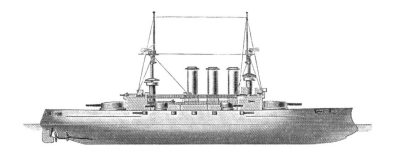

FIREFLY BOOKS

CONTENTS

HMS Victory, *see page 54*

Clermont (North River Steamboat), *see page 64*

INTRODUCTION

No one knows why our ancestors left the land and ventured out onto the water for the first time, or when it happened. They may have been hunters searching for food, or traders with goods to barter. They may have been warriors bent on conquest, or explorers longing to know what lay on the other side of a river or beyond the sea's horizon. In time, their first small primitive watercraft evolved into bigger seagoing vessels. Since then, ships and the ways in which people have used them have shaped our history, culture and civilization.

By 2550 BCE, large seagoing vessels worthy of being described as ships were being built in Egypt. One of the oldest ships to survive from the ancient world was found in Egypt in the 1950s. It was a 4,500-year-old funeral ship built for Pharaoh Khufu (Cheops). Early Egyptian ships like this were propelled by paddles or oars. The first watercraft entirely dependent on sails as a means of propulsion were developed in this region too. Dhows that have changed little for thousands of years can still be seen there today.

From simple beginnings in Egypt and elsewhere, successive generations learned how to build bigger and bigger ships, which evolved into different types of vessels specialized for different tasks. While big-bellied ships were developed for trade, slender, faster vessels were built for war. The first warships were rowed galleys; warships continued to be rowed long after cargo ships had replaced oars with sails. Before the advent of fighting at a distance with naval guns, warships had to get close enough to enemy vessels to ram them or board them. Rowing enabled warships to maneuver with greater precision than sails allowed, and to accelerate at will. The trireme, with three banks of oars, was the most successful warship in the Mediterranean for more than a thousand years. Farther north, the Vikings developed the slender, elegant longships that enabled them to raid coasts and rivers. In the Far East, the distinctive junk dominated both trade and war at sea.

Once ships were capable of oceanic voyages, explorers set out to find new lands and make their fortune, eventually circumnavigating the world and mapping many of its islands and coasts. The establishment of overseas colonies and empires brought the world's major naval powers into conflict with each other and with pirates, leading to bigger warships carrying bigger guns. The introduction of explosive shells prompted the change from wooden hulls to iron and then steel. Sometimes, great ships met an untimely end through design flaws, extreme weather or accident, as in the cases of the *Vasa*, the *Mary Rose* and the *Titanic*.

> I MUST GO DOWN TO THE SEAS AGAIN, TO THE LONELY SEA AND THE SKY, AND ALL I ASK IS A TALL SHIP AND A STAR TO STEER HER BY.
> **John Masefield, "Sea Fever"**

War, Trade, Science and Pleasure

Massive battleships reached their peak with the dreadnoughts of World War I and vessels like the *Bismarck*, *Yamato* and *Missouri* of World War II. Meanwhile, trading ships developed in two different directions. While some cargo vessels became bigger and bigger, clipper ships like the *Cutty Sark* sacrificed cargo volume for speed. When fast transport was taken over by aircraft, cargo ships continued to grow in size, to the enormous proportions of supertankers and containerships. Containerization, pioneered by ships such as *Ideal X*, transformed global cargo transportation. Meanwhile below the waves, engineers and inventors finally cracked the problems involved in traveling underwater when the first practical submarines were built at the beginning of the 20th century, changing naval warfare forever.

The desire to know more about the oceans led to expeditions concerned with scientific research. These generally used converted warships such as the *Beagle* and *Challenger*. They mapped the seabed and ocean currents, and discovered thousands of new species.

As the use of sail for commercial shipping gave way to steam, sailing for pleasure and sport grew in popularity. Winning international races like the America's Cup became a matter of national prestige, triggering an ongoing battle between the teams to harness the most advanced technology to cross the finish line first.

The desire to use aircraft in war zones where no runways on land were available led to the development of ever-bigger aircraft carriers, ultimately the giant 100,000-ton *Nimitz*-class vessels and their replacements, the equally big *Gerald R. Ford* class. These ships are nuclear-powered, giving them a virtually unlimited range.

From the moment that steam power made scheduled transatlantic journeys possible, shipbuilders created bigger, faster and more luxurious passenger liners such as the *Normandie* to operate on the lucrative route between Europe and the United States. They vied with each other to hold the Blue Riband for the fastest Atlantic crossing.

In a world where vast quantities of goods and materials are moved from suppliers to consumers, ships continue to be integral to our everyday lives, our food supplies and our security. And the reasons for going to sea have not changed since the first ships took to the sea thousands of years ago: ships still put to sea today to fish, trade, fight and explore.

LEFT: *This sketch shows the* Princeton *(1851), the U.S. Navy's first screw-propelled, steam-powered warship. She was the second U.S. warship to be named* Princeton.

BELOW: *The* Mayflower II *is a full-size replica of the famous ship, the* Mayflower, *that carried the Pilgrim Fathers to the New World in 1620.*

Pharaoh Khufu's Solar Barge

The Ancient Egyptian civilization emerged just over 5,000 years ago along the banks of the River Nile. Water transport on the river and along the coast was vital for transporting people and supplies, and also for communication and fishing. The Egyptians were quick to develop small riverboats and then larger seagoing ships for trade with Mediterranean ports. No large vessels were thought to have survived from this period, until an archaeologist made an extraordinary discovery in 1954.

TYPE: solar barge

LAUNCHED: Old Kingdom, Egypt, *ca.* 2566 BCE

LENGTH: 143 ft (43.6 m)

TONNAGE: unknown

CONSTRUCTION: Lebanon cedar planks

PROPULSION: five pairs of oars

Pharaoh Khufu, also known by his Greek name Cheops, ruled Egypt between 2589 BCE and 2566 BCE. He built the biggest of the pyramids at Giza, the Great Pyramid, but apart from that very little is known about him. The only statue of him that has ever been found is the smallest piece of Egyptian royal sculpture, a tiny seated figure just 3 inches (7.5 cm) high.

While debris was being cleared from around the base of the Great Pyramid in the early 1950s, workers discovered the remains of a wall. It was in a position that archaeologists thought was unusual: It seemed to be too close to the pyramid. One of the archaeologists, Kamal el-Mallakh, wondered if it might have been built there to hide something below it in the ground. Pits had been found elsewhere on the Great Pyramid site; they were thought to have contained some of the things the pharaoh would need in the afterlife, including ships. According to the religious beliefs of the time, the pharaoh needed a ship called a solar barge to sail the cosmic waters of the sky with the sun god, Ra. Khufu is thought to have been provided with five ships for the afterlife. Unfortunately, all the pits found until then had been emptied by robbers long ago.

When the ground was cleared, a line of huge stone blocks was found. They seemed far too big to be merely a foundation for the wall, so el-Mallakh wondered if they might be the roof of a large pit. A small test hole in the ground confirmed that the blocks covered a pit, and when el-Mallakh put his face up close to the hole, he could smell something. It was the unmistakable aroma of cedarwood,

RIGHT: *The parts of the pharaoh's solar barge were so well preserved in their burial pit that the ship could be reconstructed. Symbols carved in the pieces showed how they were intended to fit together.*

LEFT: *Khufu's solar barge was found in pieces in a pit near the base of the Great Pyramid's south side. It had lain there undisturbed for thousands of years until it was discovered in the 1950s.*

the type of wood a pharaoh's funerary ship would have been made from. At that point, Mallakh did not know what state the wood was in; it might have been devoured by ants or it might have rotted away. He called for a mirror and used it to shine sunlight into the void. He started to shake as he recognized the unmistakable shape of a ship's oar. The pit still contained the ship that had been placed there more than 4,500 years before, when the pharaoh's body was entombed in the pyramid. And it was in good condition. The ship was in pieces, an ancient flat-pack to be rebuilt in the afterlife. It was covered with the remains of ancient matting and ropes. There were 1,224 pieces, from small pegs to long planks. It took nearly 2 years to remove them from the pit and more than 10 years to conserve them and reconstruct the ship.

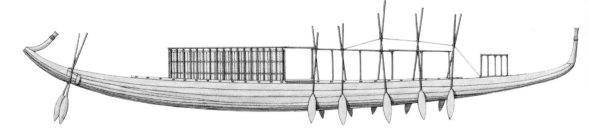

ABOVE: *The graceful, curved shape of the hull with tall stem and sternpost is described as papyriform, because it echoes the shape of earlier papyrus boats from the region.*

A Ship Fit for a Pharaoh

The hull was built using a construction method called "shell first." The hull planks were fastened together first, and then frames or struts were inserted to lock them together. Thirty planks were cut from logs as long as 76 feet (23 m) and then carefully shaped with adzes to follow the curve of the hull. The hull's shape appears to have been based on earlier papyrus-reed boats, with a tall bow and stern. Papyrus boats were common in Ancient Egypt, because there were few trees big enough to provide the wood necessary to build large ships, but papyrus reeds were plentiful. Nearly all of the wood used to build Khufu's solar barge was cedar, which would have been imported from the eastern Mediterranean. The hull had a flat bottom, with no keel; Egyptian ships would not have keels for another thousand years.

All the various pieces of the ship were fastened together by mortise and tenon joints, and tied with rope made from grass. A windowless cabin 30 feet (9 m) long sits on top of the deck. Inside this, there is a smaller chamber just long enough for a human body. Perhaps this is where the pharaoh's body lay for its final voyage. There are five oars on each side of the ship plus two steering oars at the stern. Each oar was carved from a single piece of wood.

Other pits that may once have contained boats have since been found at various tombs and temples in Egypt. Many of them were empty, but some still held the ships that had been stored in them thousands of years ago. The remains of 14 ships nearly 5,000 years old were found buried in the desert near Abydos in 1991.

THE DHOW

The dhow dominated sea travel along the coasts of Arabia, the Persian Gulf, east Africa and India up to the 15th century. It may have originated in the Mediterranean, the Indian Ocean or the Red Sea. It was developed some time between 600 BCE and 600 CE, so the last pharaohs of Ancient Egypt may have seen dhows on the Nile. Dhows range in size from small fishing boats to large seagoing vessels. Their major contribution to seafaring was the triangular "lateen" sail, which is more efficient than the square sails used in Europe and the Far East. Lateen-rigged dhows were more maneuverable, because they could sail closer to the wind than a square-rigger.

ABOVE: *Dhows were primarily trading ships with a long, slender hull and one or two "lateen" sails. The distinctive triangular fore-and-aft sails made dhows very maneuverable.*

The scarcity of trees prevented Ancient Egypt from becoming a major seafaring nation, but Khufu's solar barge is one of the oldest planked ships to survive to the present day. It is extraordinary to think that it may have carried the body of a pharaoh to its last resting place 4,500 years ago.

A Second Ship

A second pit was discovered at the base of the Great Pyramid in 1954, and it too contained a ship, untouched since the pharaoh's time. Initially, it was decided not to disturb this second ship, but tests in the 1990s revealed that the wood was deteriorating. Breaking the seal on the pit had allowed moisture and mold to enter. The ship had to be removed or it would rot and disintegrate. Radiocarbon tests on the wood confirmed it to be 4,500 years old. Once the wood has been conserved, the intention is to reconstruct the ship, but the delicate process will take several years.

LEFT: *The dismantled boat had been carefully stored in a pit carved out of the Giza bedrock and sealed with slabs of limestone.*

THE TRIREME

The Phoenicians and the Greeks, two great maritime civilizations of the ancient world, built cargo ships and warships that traveled the length and breadth of the Mediterranean. They invented the predominant warship of the era, the trireme. This oar-powered warship was made famous by the Greek navy at a crucial battle that determined the course of Western history.

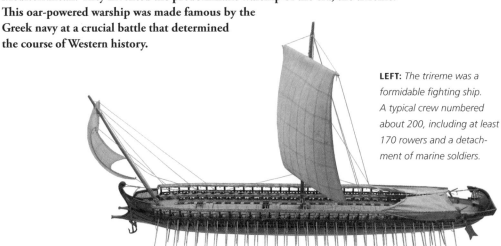

LEFT: *The trireme was a formidable fighting ship. A typical crew numbered about 200, including at least 170 rowers and a detachment of marine soldiers.*

TYPE: rowed warship

LAUNCHED: Greece, Phoenicia and Rome from about 600 BCE

LENGTH: about 120 ft (37 m)

TONNAGE: 69 long tons (70 metric tons) displacement

CONSTRUCTION: wooden planking

PROPULSION: 170 rowers and two square sails

Rowed ships called galleys had been used since the time of the pharaohs in Ancient Egypt. They had a long, slender hull and shallow draft, and were propelled principally by a bank of oarsmen on each side. They usually had at least one mast so they could switch to sail power when the wind was favorable. During times of conflict and war, ships were mainly used to transport fighting men; the ships themselves did not engage each other in combat. Then in about 800 BCE the ram was invented, and the dedicated warship was born. The ram was fitted low down on the ship's bow. A ram-armed vessel was rowed as fast as possible into the side of an enemy ship so that the ram smashed through its hull below the waterline. The Greeks developed a type of ram-armed war galley called a *penteconter*. It was propelled by 50 oarsmen, 25 on each side. A smaller version, the *triaconter*, had 30 oarsmen. In about 700 BCE the Phoenicians invented a new type of galley by adding an outrigger to the penteconter. The outrigger formed an upper gallery for a second row of oarsmen. The two rows of oarsmen gave the ship extra speed and power, and also its name — the bireme.

Then, in about 600 BCE, the Greeks or Phoenicians produced the next logical development of the bireme by adding one more row of oarsmen above the other two. This ship, the trireme, quickly became the standard warship of the eastern Mediterranean.

RIGHT: The trireme's rowers
were arranged in three
banks, or tiers. This relief
from Athens shows the top
rank of rowers in a trireme
resting their oars on the
ship's gunwale.

The Battle of Salamis

One of the most important naval battles of the ancient world
was fought between hundreds of triremes. In 482 BCE, Athenian
preparations for an expected invasion by Persia included the
construction of a fleet of trireme warships. When the invasion came
in 480 BCE, the Persians were initially victorious. They famously
annihilated a smaller Greek force on land at Thermopylae. Meanwhile,
at the naval battle of Artemisium, hundreds of Greek triremes fought
an even larger fleet of Persian triremes. The Greek fleet withdrew
when it could bear no further losses, leaving the Persians triumphant.
But the Greek commander, Themistocles, persuaded the navy to
make a stand at the island of Salamis.

RIGHT: The Battle of
Salamis was fought
between fleets of triremes
in the confined space of
the narrow straits between
the island of Salamis and
the Greek mainland.

• • • • • Greek troops

• • • • • Persian troops

‖‖‖‖‖‖ Greek fleet:

    A Corinthians
    B Athenians
    C Aeginetans
    D Other Greeks
    E Spartans

‖‖‖‖‖‖ Persian fleet:

    F Phoenicians
    G Ionians
    H Other Persians

Despite heavy losses, the Persians still had more ships, with more experienced crews, than the Greeks. They expected the Greek ships to scatter and run when they came under attack. Instead, the Greek ships used the confined space between Salamis and the Athenian coast to their advantage. They lured the Persians into the narrow straits. The Persian ships, unable to maneuver freely, became disorganized and some even rammed each other. In the ensuing battle, the Persians lost hundreds of their remaining ships and were defeated. The Greek victory at Salamis meant that the Persian army, now without effective naval support, was unable to complete its invasion of Greece.

The battle of the triremes at Salamis has assumed legendary status because of its wider significance. One analysis suggests that as ancient Greece was the cradle of Western democracy and civilization, a Persian victory at the Battle of Salamis followed by a successful invasion of Greece would have changed the whole course of Western history from then on.

The Roman Navy

The Battle of Salamis established the importance of naval power. When Rome, a mainly land-based military power, began expanding its sphere of interest in the third century BCE, it provoked hostilities with Carthage on the coast of modern Tunisia. In contrast with Rome, the Carthaginians were experienced seafarers who dominated the western Mediterranean. To defend its territory, ports and trading routes, Rome had no alternative but to build its own fleet of triremes, and even larger quinqueremes, and become a naval power itself. Quinqueremes were triremes with more than one oarsman per oar. The top two banks of oars were each rowed by two oarsmen and the lowest bank had one oarsman per oar — five oarsmen to each position, hence *quinquereme*. Each quinquereme had a crew of 400, of whom 300 were oarsmen.

The Romans invented a new tactic to gain an advantage over their enemy. They fitted their warships with a device called a *corvus* (crow). This was a bridge or gangplank that could be lowered onto an enemy ship to let Roman marines pour across and fight hand to hand. The corvus had a beak-like spike underneath that locked the two ships together.

Later, with Carthage defeated and no other naval enemy to contend with, the Roman navy evolved into a force that policed the coasts and major rivers, escorted merchant ships and suppressed piracy. Heavy

warships were unsuited to these tasks, so the navy developed smaller, lighter craft. When the Roman Empire fell in the fifth century CE, the eastern half of the empire continued for another thousand years as the Byzantine Empire, and a new warship emerged: the *dromon*. Developed from a small Roman galley called a *liburnian*, it was the principal warship of the Byzantine Empire. Dromons were fully decked ships with lateen sails. The ram in the bow had been abandoned by then. There were dromons of all sizes, with one, two and three banks of oars.

Rowed, galley-type naval vessels began to decline in the 16th century. The last major sea battle between war galleys was the Battle of Lepanto in 1571, when a Spanish-led alliance of nations defeated the Ottoman Empire. From then on, galleys were used to support larger warships under sail or for coastal and river operations, but the days of the trireme, and the many rowed ships developed from it, were finally over.

BUILDING *OLYMPIAS*

No trireme has survived from the ancient world. In 1985 a project to build a modern replica began. The ship was built in a yard in Piraeus, where many ancient Athenian triremes were built. The work took 2 years. The ship, called *Olympias*, was built from Oregon pine and Virginia oak, with a bronze-covered ram on its bow. Bracing ropes helped to stop the hull from flexing (called "hogging" and "sagging") as the ship passed over waves. In sea trials, *Olympias* exceeded all expectations. With a full complement of 170 oarsmen, she achieved a speed of 9 knots (10 mph or 17 km/h) and made a 180-degree turn within 1 minute, in only two and a half ship-lengths. *Olympias* was then commissioned as a warship in the Greek navy, the only warship of this type in any modern navy.

ABOVE: Olympias *is a full-size replica of an Athenian trireme from the fifth and fourth centuries* BCE. *Launched in 1987, the 77-short-ton (70-metric-ton) vessel is the only one of its kind.*

THE NYDAM SHIP

When a Danish teacher went digging in a local bog in the middle of the 19th century, he found the oldest-known rowed vessel in northern Europe. The Nydam ship, as it is known, had been submerged in the bog for more than 1,600 years, and is still one of the finest examples of a pre-Viking Nordic vessel. It is significant for being one of the evolutionary steps in early Nordic ship design that led to the Viking longship.

TYPE: pre-Viking warship

LAUNCHED: Southern Denmark, ca. 315 CE

LENGTH: 75 ft (23 m)

TONNAGE: Just over 3 long tons (3 metric tons) displacement

CONSTRUCTION: oak, clinker-built (lapstrake)

PROPULSION: 15 pairs of oars

BELOW: *The Nydam ship is the oldest-known rowed and clinker-built vessel from northern Europe. It lay submerged in a bog, undiscovered, for more than 1,600 years.*

In 1851, Denmark won the First Schleswig War against a German alliance of Prussia, Schleswig and Holstein. When the Danes occupied parts of Schleswig-Holstein, they removed officials and others who had sympathized with the enemy. This created vacancies for teachers. One of these new teachers was Conrad Engelhardt (1825–81). His duties included taking responsibility for a local cache of ancient finds called the Jaspersen Collection. He asked local farmers and peat cutters to add to the collection by donating some of the swords, pots and other objects they had found. Engelhardt himself also went digging to search for more items.

In 1859, he started digging in an area called Nydam Mose (Nydam Bog), about 5 miles (8 km) from Sønderborg in southern Denmark. Over the next 4 years, he found three ancient boats and numerous other objects dating from the Iron Age, including more than a hundred swords. At this point, war intervened again in the shape of the Second Schleswig War (1863–64). This time Denmark lost, and ownership of Engelhardt's discoveries passed to Prussia. Unfortunately, one of the boats he had found was destroyed and lost forever when troops used it as firewood during the war. But the Nydam ship survived and was taken to Kiel, the capital of Schleswig-Holstein. Later it was moved to the Archäologisches Landesmuseum at Gottorf Castle in Schleswig, where it is still on display today.

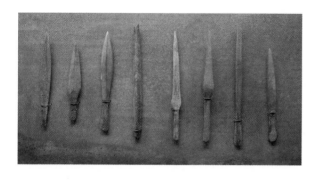

The Nydam ship is made of oak. Its keel is a single piece of oak 47 feet (14.3 m) long and 22 inches (57 cm) wide. There are five planks or strakes on each side. These and the stem and sternpost are fixed together and to internal frames with wooden pegs and iron rivets. Propulsion was by means of oars, 15 on each side. The oars were no more than 11 ft 6 in (3.5 m) long, so the rowing action must have involved short, quick strokes. In 1993, a broad steering oar was found. It was used as a side rudder, fixed to the starboard side of the stern. Inside the ship, there is no evidence of a mast or a mast foot, a block of wood that would have supported a mast, so the Nydam ship probably never had a sail. It would be another 350 years before Scandinavians mastered the use of sails. The ship's long, narrow hull shape and the number of weapons found with it indicate that it was a fighting ship, an ancient warship. It represents the state of ship technology just before the development of the Viking longship.

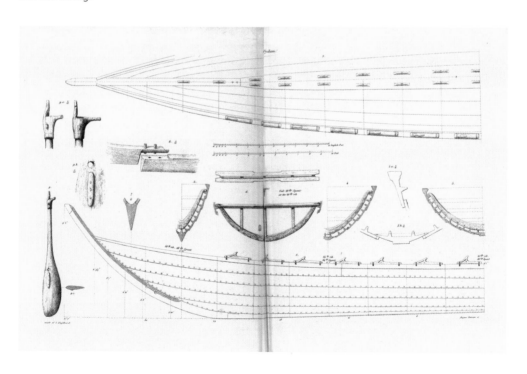

A Sacred Site

Later studies of the Nydam Bog revealed that objects had been deposited there on up to eight separate occasions. The four biggest depositions, known as Nydam A to D, were made between about 200 and 400 CE. At that time, the bog would have been a lake. The Nydam ship, the biggest of the three vessels found by Engelhardt, has been dated to the third major deposition in 350 CE. Dendrochronology (tree-ring analysis) dated the wood it was made from to about 310–20 CE. There were at least three more depositions up to about 475 CE. The last involved the placement of about a thousand objects surrounded by a "fence" of 36 swords.

Some of the best-known ancient ships found in northern Europe and Scandinavia were buried as part of a grave, either as grave goods (items buried with a body) or as a resting place for one or more bodies. The Nydam ship is different. It appears to have been buried along with the other boats and weapons as spoils of war. They were probably seized from a defeated enemy after a battle and offered to the gods in gratitude for the victory.

So many weapons were found in the bog that they revealed something of the organizational structure of Iron Age armies. Most of the warriors were armed with lances or spears. Most of the rest were armed with swords and the remainder were archers and cavalry.

After some new objects were discovered in 1984, the Institute of Maritime Archaeology at the Danish National Museum undertook a detailed survey of the bog. They unearthed more objects including parts of the Nydam ship that Engelhardt had missed. These included two carved wooden heads that are thought to have fitted on either side of the ship's bow as mooring posts. In 2011, a fourth boat was found. It was even bigger than the Nydam ship, but it was too badly damaged to be reconstructed.

ABOVE: *Conrad Engelhardt's excavations in the Nydam bog and other locations in Denmark produced an assortment of personal possessions from the Iron Age, including weapons, clothing, combs, pendants, earrings and buttons.*

RIGHT: *The Nydam ship was steered by means of a side-mounted oar, which was the usual steering mechanism in northern Europe until the rudder was adopted in the late Middle Ages.*

The Evolution of the Longship

The Nydam ship is one of a series of discoveries that reveal the evolution of the longship. In 1920, archaeologists first learned of a boat that had been unearthed decades previously, in 1880, by farmers cutting peat at the Hjortspring Farm on the Danish island of Als. The Hjortspring boat was originally about 62 feet (19 m) long, of which 46 feet (14 m) had survived. Dated to about 350 BCE, it is a relatively simple paddle-powered "war canoe" built by sewing planks together. The planks overlap, showing the "clinker" or "lapstrake" construction used in many later ships.

The Nydam ship, dated to about 315 CE, is a later, bigger and more advanced design. Its longest planks, or strakes, were each made from two or more pieces of timber expertly connected by invisible joints. The riveted planks and the use of oars are an advance on the Hjortspring boat.

Then, in 1920, two seventh-century ships were found buried in Kvalsund, Norway. The larger of the two was wider than the earlier vessels and had a keel, enabling it to hold its course better in bad weather. This Kvalsund ship had most of the characteristics of a Viking longship. The next step was to be the longship itself.

SUTTON HOO

Ship burials, sacrifices and offerings represent an important record of ancient ship design, and not just in Scandinavia. In 1939, Mrs. Edith Pretty asked the archaeologist Basil Brown to investigate a large Anglo-Saxon burial mound on her land at Sutton Hoo in Suffolk, England. Inside the mound, Brown discovered the imprint of an 89-foot (27-m) ship, probably dating from the early seventh century CE. The ship had rotted away, but its shape was clearly visible in the soil. The iron rivets that fastened its planking together were found. It housed a burial chamber packed with spectacular treasure including silverware, jewelry, an iron helmet and the remains of clothes.

ABOVE: *None of the Sutton Hoo ship's timber survived, but it left the unmistakable imprint of a ship in the ground.*

LEFT: *One of the most important finds from Sutton Hoo is a magnificent, decorated Anglo-Saxon helmet.*

ISIS

While Scandinavian boat builders were progressing toward the development of the longship, the Roman Empire was spreading across Europe. The growth in Roman trade required merchant ships. In 1988, undersea explorer Robert Ballard made a unique find: an untouched fourth-century Roman merchant ship, too deep to have been plundered by divers. Ballard used *Isis* to pioneer a new way of sharing the exploration of a shipwreck with students elsewhere in the world, by means of robot submersibles and satellite links.

TYPE: merchant ship

LAUNCHED: Roman Empire, 4th century CE

LENGTH: about 100 ft (30 m)

TONNAGE: 100–200 tons

CONSTRUCTION: lead-covered wood (white oak)

PROPULSION: oars and square sails on two masts

*R*obert Ballard and his team were scanning the bottom of the Mediterranean Sea for signs of ancient shipwrecks. As his ship, *Starella*, sailed over a muddy underwater plateau called the Skerki Bank between Sicily and Tunisia, a camera sled called *Argo* towed by the ship relayed live pictures to the surface. The Skerki Bank lay 2,500 feet (760 m) below the sea route between Carthage and Rome. If an ancient ship had been overwhelmed by bad weather here, its remains could be lying on the Skerki Bank.

On June 3, 1988, something appeared on the screens in the control cabin on board *Starella*. It was the unmistakable shape of an amphora, a storage jar. Roman cargo ships were loaded with thousands of them, used to transport olive oil, fish and wine. Then *Argo*'s cameras found more amphoras; Ballard's team had found the debris from an ancient Roman ship. They began a systematic search for the ship from which the amphoras had fallen.

On June 12, the observers in the control cabin suddenly saw a large pile of amphoras and other objects on their screens. It appeared to be the place where the ship had come to rest on the seabed. They named her *Isis*. There was no sign of the hull or rigging. Any sacks of grain the ship might have been carrying were gone too, all eaten by sea creatures or broken down long ago. But a pile of amphoras, a jug, a pot and a grindstone were clearly visible. Some metal remains might have been part of a stove or possibly the ship's anchor. It was the first time anyone had laid eyes on these objects since the ship sank.

Ballard wound up his operations for that year and planned his return the following year for a more ambitious project. He had pioneered a technique called "telepresence" to beam live pictures from the seabed to classrooms thousands of miles away where children could watch the pictures and

ABOVE: *Round-hulled cargo ships like this were the workhorses of the Roman Empire, constantly plying the Mediterranean trade routes with essential supplies. They could transport goods and materials faster than taking them overland.*

LEFT: *The Roman cargo ship discovered by Dr. Robert Ballard was found on the Skerki Bank. It lies between Sicily, Sardinia and Tunisia, beneath a busy ancient sea route between Carthage and the Roman port of Ostia.*

ask the team questions. He had established the Jason Project to use this technology to inspire students to take up careers in science, technology, engineering and mathematics. The 1989 expedition would be its first outing.

Now based on the research vessel *Star Hercules*, Ballard's team was using a new camera sled called *Medea* and a more advanced ROV (remotely operated vehicle) called *Jason*. The project's first transmissions explored an active underwater volcano, the Marsili Seamount, at a depth of 1,000 feet (300 m). Then it was time to find *Isis*. Just as the next live broadcast began, *Medea* homed in on the wreck. They mapped the whole site in detail. Accurate mapping was vital, because information about where objects were found could be useful to archaeologists now and in the future. The size of the debris field and the types of objects found convinced the team that this was the wreck of a Roman cargo ship.

A typical Roman cargo ship had a high bow and stern. The sternpost curved forward, often in the shape of a swan's neck. There were two masts. The main mast, amidships, carried a square sail, sometimes with an additional triangular topsail above it. A smaller mast at the front leaned forward over the bow and carried a smaller sail called the *artemon*. Small cargo ships like *Isis* carried up to 3,000 amphoras; bigger vessels could carry 10,000.

LEFT: *This fragment of a painting from Ostia shows a cargo ship called* Isis Giminiana *being loaded with grain. The* magister *(captain) is named Farnaces and the ship's owner is Arascantus.*

To the Surface

With the mapping complete, *Jason* was sent down to start lifting objects. Experts doubted whether amphoras could be raised from such a great depth (2,684 feet or 818 m) without falling apart. *Jason*'s mechanical arm lifted each amphora and placed it in a net bag, which was part of a device called an elevator that sat on the seabed. An acoustic signal sent down through the water made the elevator release a heavy steel weight. Floats attached to the elevator were then buoyant enough to lift it to the surface, where its precious cargo was transferred to the mothership, *Star Hercules*.

The amphoras survived their journey to the surface. They were of a type made in Tunisia, so the ship appeared to have been sailing from North Africa toward Italy. The style of the amphoras indicated that the ship probably sank in the third or fourth century CE. One of the objects recovered was an oil lamp. This was a very useful find, because Roman oil lamps can be dated quite accurately. The lamp from *Isis* dated from

RIGHT: *Dr. Robert Ballard addresses students from his research vessel. Ballard pioneered the use of satellite technology to connect students, teachers and scientists with remote undersea wreck sites in real time, a phenomenon known as telepresence.*

NUCLEAR-POWERED DIVER

NR-1 was a nuclear-powered research submarine. It was launched in 1969 and remained in service until 2008. Displacing 400 long tons (406 metric tons), it could explore the oceans to a depth of 3,000 feet (almost 1,000 m), giving access to most of the world's continental shelves. As nuclear submarines go, NR-1 was small, only 145 feet (44 m) long. Unlike most submarines, it had three viewing ports and exterior lighting so the crew could look outside. And, uniquely, it had retractable wheels underneath so it could crawl along the seabed. In 1997, NR-1 was used in a wide-area search to identify promising wreck sites, which were then explored in detail by the Jason ROV.

BELOW: *Submarines normally avoid contact with the seabed, but* NR-1 *was designed to drive along it! Its nuclear powerplant also gave it greater submerged endurance than any other research submarine.*

the second half of the fourth century. Then, when a pot recovered from the wreck site was x-rayed, a coin was spotted at the bottom. It turned out to be a bronze *centenionalis* minted in about 355 CE, confirming the oil-lamp evidence.

Undiscovered Treasures

Ballard's team was only equipped to collect objects lying on the seabed, but they suspected that the rest of the ship still lay under the mud, perhaps with thousands more amphoras. A small part of the ship's hull was found and recovered. It showed the mortise and tenon construction that was typical of a Roman ship. It would have been built shell-first: hull planking first, joined together by mortise and tenon, followed by internal frames and cross-members.

Ballard returned to the *Isis* wreck site in 1997 with a new vessel, a small nuclear-powered submarine called *NR-1*. Its advantage was that it could stay submerged for up to 30 days. Using *NR-1*, another five ancient wrecks were discovered.

MORA

The Norman conquest of England was one of the most influential events of the past thousand years. It's almost impossible to make sense of the past millennium of British history without taking into account the ambitions of the Norman nobility who took control of England in 1066. From the birth of the English language as we know it to the spread of Protestantism and the colonization of whole continents, its impacts are diverse and far-reaching. And it began with an invasion fleet led by William of Normandy's flagship, *Mora*.

TYPE: *drakkar* longship

LAUNCHED: Barfleur, France, 1066

LENGTH: unknown

TONNAGE: unknown

CONSTRUCTION: wood, lapstrake-planked

PROPULSION: one square sail, 60–70 oars

When England's King Edward "the Confessor" died without an obvious heir in January 1066, the door was open to conflicting claims of succession. As Edward's brother-in-law and one of England's most powerful nobles, Harold Godwinson, Earl of Wessex, was made king (Harold II), but he was immediately challenged by his brother Tostig, by Norway's King Harald Hardrada and by Edward's cousin William, Duke of Normandy. Although Harold expected William to launch an invasion across the Channel, Tostig and Harald struck first. Harold was compelled to march his armies more than 240 miles (390 km) to Yorkshire, where he defeated the combined forces of Tostig and Harald at the Battle of Stamford Bridge. Then he had to quick-march his armies back south to meet William.

William, meanwhile, had assembled an invasion force of perhaps 7,000 men, including cavalry and archers. And unlike Harold's forces, they were fresh and ready for battle.

LEFT: *Following the Battle of Stamford Bridge, where he defeated an invading Viking army, King Harold marched his army 250 miles (400 km) south to Hastings in just 2 weeks to meet a second invasion, this time from Normandy.*

→ William the Conqueror's campaigns, 1066

→ Harold's route from York to Hastings

The Invasion Fleet

The size of William's fleet is unknown, but it probably numbered in the hundreds of ships. The types of ships he used are clear. William's domain was the region of France that had recently become known as Normandy, because it was settled by the Northmen — that is, Norsemen. And although these descendants of Vikings had largely adopted the French language, they still maintained their traditional shipbuilding methods.

Some men and gear must have been carried in *knarrs*, the cargo ships that the Norsemen had used to settle Iceland and Greenland in the 9th and 10th centuries. But William's armada relied primarily on longships — warships designed specifically for amphibious assaults. Light in weight and of shallow draft, longships could be landed and hauled up on any gently sloping beach. Since an invading force is nowhere so vulnerable as when landing, the speed with which the longship could be disembarked and unloaded made it a formidable weapon of war. Vikings had proved this again and again, using longships on successful raids and invasions of lands from Ireland to Ukraine.

Longships appeared in many varieties, distinguished mostly by their length and number of oars, but the largest and fastest type was the *drakkar*, or dragon ship.

William's Flagship

As a claimant to the throne of England, William needed a ship of great prestige with which to lead his expedition. His wife, Matilda of Flanders, therefore had the *Mora* built for him in Barfleur, a drakkar which was the largest ship in William's fleet. Over 100 feet (30 m) in length, *Mora* may have had as many as 34 rowing positions on each side, meaning a minimum of 68 oarsmen, or possibly twice that, if two men pulled each oar. In addition to William's own retinue, she carried 10 knights, which doesn't sound much until one realizes that each knight was accompanied by several retainers, a horse or two and a mountain of personal cargo to equip and sustain him.

Mora was beautiful. She had a multicolored sail and a gilt figure-head of a child blowing an ivory trumpet, pointing forward — toward England, presumably.

Longships

Mora was a longship par excellence: larger, richer and faster than most, but in most respects squarely in the mold of warships of the Viking age. A result of centuries of development, longships were narrow, double-ended (pointed at both bow and stern) and lapstrake-planked. They were easily driven under oars and exceptionally fast when powered by their large, single square sail, sometimes planing or skimming over the surface of the water at speeds of 15 knots (17 mph or 28 km/h). Built with true keels, which allowed them to sail upwind, they were nonetheless flexible enough to "give" with the waves.

The gunwales were pierced by narrow oar-ports to bring the oars down to an efficient height above the waterline. When heeling under sail, the oar-ports were covered by hinged shutters to keep water out. There was no cabin or other superstructure, not even rowers' benches. It is assumed that the oarsmen sat on their own sea chests. A large, single steering oar, or "steer-board," hung on the boat's right quarter. Its position is the origin of the word "starboard."

The longship's light weight and shallow draft meant it could be sailed and rowed far up rivers and man-hauled around otherwise impassable shallows and rapids. And the Vikings used this to good effect. Starting in the ninth century, they began ranging far and wide, making lightning raids in Scotland, Wales and Ireland, along the entire northern coast of Europe and even to the shores of southern Spain and northern Africa. Even before the Norman Conquest, they had invaded and settled in

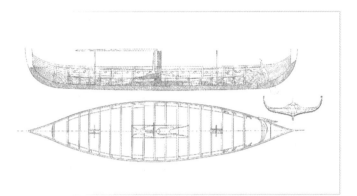

ABOVE: *With their light weight and shallow draft, longships could be beached, making amphibious attacks quick and deadly. They were the vessel of choice of both of Harold's main opponents.*

VIKING SHIP FINDS

Several archaeological finds have given us an excellent understanding of Viking ship construction methods. The most important are the Oseberg and Gokstad ships, both used in high-status burials in the county of Vestfold, Norway, and now on display at the Viking Ship Museum in Oslo. Five other Viking ships that were deliberately sunk to protect the harbor of Skuldelev, Denmark, from invaders are now displayed at the Viking Ship Museum in Roskilde, Denmark.

RIGHT: *Researchers have learned much about the sailing characteristics of Viking-age vessels from modern reconstructions.*

England, establishing the Danelaw that stretched from London to Yorkshire. Traveling inland along the Dnieper and Volga rivers, they founded the country that would become Russia, and even settled in southern Italy, Sicily and Constantinople.

The Invasion

William launched his attack late on September 27, while Harold's forces were still away in the north. A lantern was hoisted to *Mora*'s masthead, and a horn was blown regularly, so that the fleet could follow in the dark. Even so, *Mora* was so fast that, at daybreak on the 28th, the rest of the fleet was nowhere in sight. *Mora* dropped her sail to allow the fleet to catch up. While waiting in the middle of the English Channel, the chronicles say, William breakfasted "with wine."

William's landing in Pevensey, on the south coast of England, was unopposed, and he was able to consolidate his position and establish strong defenses by the time Harold's exhausted army met him to do battle near Hastings on October 14. The battle raged back and forth until Harold was killed. It was the decisive event in the Norman victory that transformed William into "The Conqueror."

BELOW: *Part of the Bayeux Tapestry depicts William the Conqueror's invasion fleet, including his own ship,* Mora, *crossing the sea to England.*

ZHENG HE'S TREASURE SHIPS

For much of its 2,000-year imperial history, China shut itself away from the rest of the world and looked inward. But in the early 15th century, China sent seven great fleets of ships out into the Indian Ocean as far away as Africa. If contemporary accounts can be believed, some of these ships were giants, bigger than anything that had been built up to then or would be built anywhere for hundreds of years. All seven fleets were commanded by a legendary eunuch admiral called Zheng He.

TYPE: junk

LAUNCHED: China, early 1400s

LENGTH: up to 538 feet (164 m) (disputed)

TONNAGE: 19,700–29,500 long tons (20,000–30,000 metric tons) displacement

CONSTRUCTION: wood planking

PROPULSION: 12 square-rigged sails on 9 masts

The third Ming emperor of China, Zhu Di (also known as the Yongle Emperor), was particularly aggressive in his defense of China against all possible enemies. He himself led military campaigns against Mongolian tribes and sent armies to deal with Manchurian, Korean, Vietnamese and Japanese threats. To deter potential enemies farther afield, he chose a trusted general called Zheng He. In 1403, the emperor ordered the construction of a *Xiafan Guanjun* (foreign expeditionary fleet) that Zheng He would command.

The fleet was vast in scale. It included more than 60 giant treasure ships called *baoshan*. They were said to be up to 540 feet (164 m) long. They had four decks, a double hull and watertight compartments. They were reputed to have a displacement of up to 29,500 long tons (30,000 metric tons) and were capable of carrying 2,500 (2,540 metric tons) of cargo. If this is true, they would be the biggest wooden ships ever built, and easily double the length of the biggest ships in Europe at that time. When wooden ships approaching this size were built in Europe in the 19th century, they suffered from structural problems that required iron bracing to stop their hulls from breaking up.

Chinese naval technology in the 15th century was far more advanced than in Europe. Even so, reports of the giant Chinese treasure ships were assumed to be exaggerations by

RIGHT: *This bronze statue of the Yongle Emperor, who ordered the construction of Admiral Zheng He's treasure fleet, shows what an imposing character he was. The statue stands in the Ming Dynasty Tombs in Beijing, China.*

RIGHT: *This full-size replica of a middle-sized Zheng He treasure ship stands at the site of the Treasure Boat Shipyard in Nanjing, China. It was built from concrete with wooden planking.*

writers with overactive imaginations. Then in 1962 the rudderpost of a treasure ship was found at an archaeological dig in Nanjing, an inland port that served as China's capital in the early Ming period (1368–1421). It was 36 feet (11 m) long. Judging by the lengths of rudderposts on known ships, the ship this rudderpost was made for is likely to have been about 500 feet (150 m) long.

Zheng He's Seven Voyages

The first treasure fleet was the biggest. Up to 60 treasure ships set sail from Nanjing in 1405, accompanied by hundreds of warships and cargo ships full of food and other supplies. The warships were among the smallest and lightest ships in the fleet for maximum maneuverability in battle, but even these were twice as long as Christopher Columbus's flagship, *Santa María*. The treasure ships were up to five times the length of *Santa María*.

A total of 317 ships carrying more than 27,000 men sailed to Vietnam, Siam (Thailand today), Java, the Straits of Malacca, Cochin and finally Calicut (now Kolkata) on the west coast of India. The treasure ships were packed with fine porcelain, silk, tea and lacquerware as gifts for foreign heads of state. In return, the treasure fleets received gifts of spices, ivory, pearls, gemstones and medicines. The fleet returned to China in 1407. Smaller fleets in 1407 and 1409 repeated the same voyage. A fourth expedition in 1413 went farther, reaching Hormuz in the Persian Gulf. The fifth and sixth expeditions in 1417 and 1421 returned heads of state and ambassadors to their home countries. The fifth voyage went as far as the east coast of Africa, bringing back exotic

animals including lions and zebras that had never been seen before in China. The seventh and final voyage in 1431 sent ships into the Red Sea. Wherever the fleets found war, rebellion or pirate activity, they quashed it. In 1433, while the seventh fleet was returning to China, Zheng He died. The 62-year-old admiral was buried at sea. It is said that a braid of his hair was brought back to China for burial.

Ming Shock and Awe

These grand voyages were not primarily intended for exploration or trade. Chinese merchants had already ventured as far as India and east Africa. They were part of the emperor's determination to intimidate all potential enemies by demonstrating China's immense power. This was Ming dynasty "shock and awe." The message was: "Look at what we can do, the mighty ships we can build and the advanced technology we have at our disposal. Don't mess with us, or else!" An ulterior motive for the expeditions was to search other countries for any sign of the Yongle Emperor's predecessor, the Jianwen Emperor, who had disappeared in mysterious circumstances. When his palace burned down, his body could not be identified among the dead and there were rumors that he had escaped in disguise. The Yongle Emperor feared his return to seize the throne, but no trace of him was ever found.

Then, quite suddenly, China turned its back on seaborne power projection. There were no more treasure fleets, because there were greater priorities at home and also a shortage of money. The Yongle Emperor had strained the state treasury by moving the country's capital from Nanjing to Beijing and constructing the vast Forbidden City there as his new base. In addition, natural disasters had caused devastating famines, floods and epidemics. Hostile tribes on China's land borders were becoming increasingly aggressive, too.

The Yongle Emperor died in 1424. He was succeeded by his son, the Hongxi Emperor, who ordered an end to the treasure fleets. His reign lasted only 9 months. He was succeeded by his son, the Xuande Emperor, who allowed Zheng He to make one more expedition, the seventh. It was finally decided that the treasure fleets had been of little value to the state considering their great expense, and there were to be no more. The vast fleet of ships, including the giant treasure ships, were either burned or left to rot. All overseas trade was banned and it was made a capital offense to sail from China in a multimasted ship. The navy was allowed to shrink to a fraction of its former size and eventually all large ships were ordered to be destroyed. Confucian courtiers and scholars, who had been opposed to the treasure fleets, did their best to eradicate all evidence of their existence from official records so that later emperors would not be tempted to duplicate them. It is interesting to speculate that if the treasure fleets had continued and had ventured even farther, it could have been China, not the European powers, that colonized far-flung parts of the world over the next couple of centuries.

BELOW: *ale model of a turtle s... ands on display at Kor... War Memorial museu... Seoul, South Korea... armor-plated deck and t... wooden hull prote... the crew inside.*

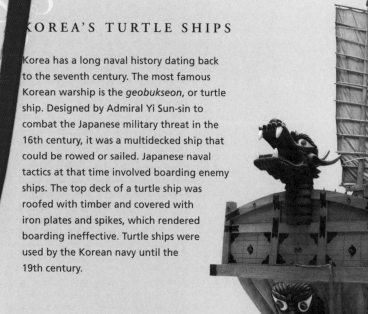

KOREA'S TURTLE SHIPS

Korea has a long naval history dating back to the seventh century. The most famous Korean warship is the *geobukseon*, or turtle ship. Designed by Admiral Yi Sun-sin to combat the Japanese military threat in the 16th century, it was a multidecked ship that could be rowed or sailed. Japanese naval tactics at that time involved boarding enemy ships. The top deck of a turtle ship was roofed with timber and covered with iron plates and spikes, which rendered boarding ineffective. Turtle ships were used by the Korean navy until the 19th century.

SANTA MARÍA

The discovery of the New World by Europeans, and its subsequent settlement, is one of the most significant events in world history. Although Christopher Columbus was not the first explorer from the Old World to reach the Americas, it was Columbus's voyages that led to further exploration and settlement.

TYPE: *nao* (carrack)

LAUNCHED: Pontevedra, Galicia, 1460

LENGTH: approx. 62 ft (19 m)

TONNAGE: approx. 108 tons

CONSTRUCTION: wooden hull

PROPULSION: five sails on three masts and bowsprit

Christopher Columbus, an experienced merchant seaman born in 1451, was convinced that it was possible to sail to China and India by heading west, but he needed help from a wealthy patron to prove it. After failing to find support in France, England and Portugal, he tried Spain. King Ferdinand and Queen Isabella of Castile and Aragon, keen to secure overseas territories, decided to back him. They provided him with three ships: *Santa María*, *Niña* (*The Girl*) and *Pinta* (*The Painted One*). The biggest and slowest of the three, *Santa María*, was captained by Columbus, who called her *La Capitana* (the flagship). She was a three-masted carrack, or *nao* in Spanish (*nau* in Portuguese), with a crew of about 40. The *Niña* and *Pinta* were smaller caravels, also three-masters, with crews of about 20.

The voyage Columbus was proposing to make was extraordinary for the end of the 15th century. At that time, ships rarely ventured more than a few hundred miles on the open sea. Columbus was about to embark on a voyage of several thousand miles. He based his ability to make the voyage and return alive on three mistaken beliefs. He thought the world was significantly smaller than it actually is; he thought the Asian landmass extended farther east than it does; and he thought he would be able to break his journey at an island called Cipangu, believed to be about 1,500 miles (2,400 km) east of Asia.

CARRACKS AND CARAVELS

The carrack was a popular cargo ship in the 15th century. It carried square sails on the foremast and mainmast, and a lateen sail on the mizzenmast. Square sails were used for speed and the lateen sail aided maneuverability. A square sail could also be carried under the bowsprit. The caravel was a smaller ship, originally a fishing vessel, but developed later into a bigger cargo carrier. With a higher length-to-beam ratio, caravels were more controllable and maneuverable than carracks. Originally lateen-rigged (*caravela latina*), they could also be rigged with a mixture of square and lateen sails (*caravela redonda*), making them faster when running before the wind. The bow and stern of both carracks and caravels were raised to form a forecastle and sterncastle.

ABOVE: *Christopher Columbus's tiny fleet of ships begin their historic voyage across the Atlantic Ocean to the New World.* Santa Maria, *the biggest of the three vessels, served as Columbus's flagship.*

Searching For Asia

Columbus's three little ships set sail from Palos in southern Spain on August 3, 1492. They stopped at La Gomera in the Canary Islands for final repairs and supplies for the ocean voyage. The *Niña*'s rig was also changed from lateen to a mixture of square sails and lateen, transforming her into the fastest vessel in the little fleet. They departed from La Gomera on September 6. A month later, with the crews close to mutiny and Columbus on the point of turning back, they started finding pieces of wood and vegetation floating in the sea and saw birds in the sky. They rightly assumed that they must be close to land. On October 12 they finally spotted land in the distance. As they approached it they could see that it was a small island and there were people on the shore. Columbus claimed the land for Spain and named it San Salvador. Its precise identity and location are uncertain, but it lay in the island group known today as the Bahamas. The expedition's goal was to reach Asia, so the ships sailed on. Seven local men they took with them as guides led the ships to Cuba, which Columbus mistakenly thought was a peninsula of Asia. He also saw natives smoking tobacco for the first time.

THE SEA WILL GRANT EACH MAN NEW HOPE,
AS SLEEP BRINGS DREAMS OF HOME.
Christopher Columbus

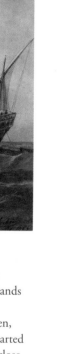

SANTA MARÍA

THE VIKINGS
IN AMERICA

Christopher Columbus is celebrated as the first European to visit the New World, but Viking explorers got there first. Five hundred years before Columbus reached the Caribbean, a Viking called Leif Eriksson, son of Erik the Red, landed in Newfoundland. The accounts of his voyage that were passed down the generations by word of mouth differ. One tells of him going off course while sailing from Iceland to Norway and accidentally landing in North America. Another story claims that his historic voyage was no accident; he was said to have been retracing the steps of an Icelandic trader, Bjarni Herjólfsson, who had seen the North American coast a decade earlier. Eriksson called the land he found Vinland (Wineland), because he found grapes perfect for making wine. He spent one winter in Vinland and then returned home. Archaeologists found a ruined Viking-style settlement at L'Anse aux Meadows in northern Newfoundland in the 1960s.

Pinta's captain set off on his own, without Columbus's permission, to explore independently. On December 5, the other two ships arrived at Hispaniola (now Haiti and the Dominican Republic). As *Santa María* was sailing along the coast of Hispaniola on Christmas Eve, Columbus went to his cabin to get some sleep for the first time in 2 days. The weather was so calm that the sailor tasked with steering the ship handed over control to a cabin boy.

At midnight, the ship ran aground on a sandbank. Columbus sent the ship's master away in a boat to lay an anchor astern in an attempt to pull the ship off the bank. Instead, the master tried to escape in the boat to the *Niña*, but the crew refused to let him aboard. Columbus ordered *Santa María*'s masts to be cut away to lighten the load in an attempt to float off the bank. But then she fell over on her side and her hull timbers split open. The ship was lost.

The *Niña* was too small to carry both crews, so 39 sailors were left behind on Hispaniola. The town they founded, Villa de Navidad (Christmas Town), was the first European settlement in the New World. The *Niña* set sail on her homeward journey at the beginning of January 1493. A few days later, *Pinta* rejoined her. The two ships became separated during a storm, but both survived. After a brief stop in the Azores in the middle of February, Columbus continued eastward. Another storm forced him into Lisbon, Portugal. He finally reached Spain on March 15.

LEFT: *The Viking explorer Leif Eriksson may have been the first European to set foot on the North American mainland. Leif Eriksson Day, October 9, commemorates the event in the United States.*

A Hero in Chains

Columbus would make three more voyages to the
New World, exploring the Caribbean and the coast
of South America. His second fleet included Spanish
settlers. When he returned to Hispaniola to rescue
the crewmen he had left behind, he found that they
had been killed by natives. Columbus began to fall
from favor toward the end of his third voyage.
Judged to have mismanaged the new colonies he had
established, he was brought back to Spain in chains
and his property was seized.

ABOVE: *This painting by
Sebastiano del Piombo (also
known as Sebastiano Luciani)
is said to depict Christopher
Columbus, although it was
painted several years after
his death.*

He eventually regained his freedom and, although in poor health,
embarked on a fourth voyage. His ships leaked so badly that he had
to beach them on the coast of Jamaica. Columbus and his crew were
marooned there for a year before they were rescued. His health
deteriorated over the next year and a half until his death on May 20,
1506, at his home in Valladolid in Spain.

RIGHT: *Columbus's first
voyage to the New World
heralded a new age of
European exploration and
conquest. He always insisted
that he had found a route
to Asia and never accepted
that he had discovered
another continent.*

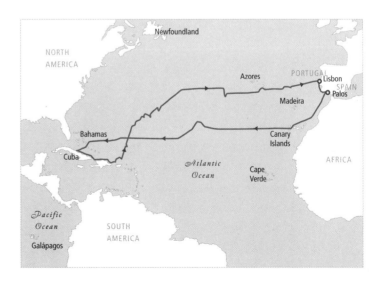

MARY ROSE

The Royal Navy warship *Mary Rose* is one of the most famous ships of the early modern era. Its fame arises not from its achievements in service, but from the mystery surrounding its sinking and from its return to dry land after more than 400 years on the seabed. It was also one of the first warships to incorporate the newly invented gunport. The thousands of items found in its wreck gave archaeologists an unparalleled insight into life aboard a Tudor warship.

TYPE: carrack-type warship

LAUNCHED: Portsmouth, England, 1511

LENGTH: 105 ft (32 m) keel

TONNAGE: 500 tons initially, 700–800 after refit

CONSTRUCTION: clinker-built (lapstrake) initially, later converted to carvel (flush) hull planking

PROPULSION: 9–10 sails on four masts and bowsprit

BELOW: *King Henry VIII expanded the fledgling "Navy Royal" to defend his realm. The* Mary Rose *was one of the first "great ships" he ordered.*

Faced with the growing sea power of England's enemies, principally France and Spain, King Henry VIII laid the foundations of the modern Royal Navy by building a fleet of warships to defend his realm. One of these was the *Mary Rose*, one of the biggest ships in the navy. Her keel was laid in 1510 and she was launched the following year. She served in two wars against France and was then held in reserve until 1535.

At about this time, she underwent a major refit. Her hull was converted from clinker to carvel planking. The original overlapping planks were removed, then the notched frames behind them were adzed smooth so that flush "carvel" planking could be fitted. Gunports were cut in her sides. The introduction of gunports was made necessary by the growing size and weight of naval guns. To maintain a warship's stability, the heavy guns had to be mounted lower in the hull and so openings had to be made in the ship's sides to fire through. As the gunports were close to the waterline, they had to be capable of being sealed when necessary. Smooth carvel planking made it possible to fit the gunports with watertight lids.

The French Attack

Following Henry's break with the Roman Church and establishment of the Church of England, the king feared attack and invasion by one of Europe's Catholic powers, especially England's long-term enemy, France. On July 16, 1545, more than 100 French ships entered the Solent, the strait that separates the Isle of Wight from the English mainland. About 80 English ships were waiting for them. One of them was the newly refitted *Mary Rose*.

The two fleets met and fought on July 18. There were few losses on either side in the initial engagement. The next day, there was so little wind that the sailing ships could not maneuver, but the French had brought rowed galleys with them. These headed for the becalmed English fleet and opened fire. Then the wind picked up and the English ships were finally able to get under way. Two of the biggest ships, the *Henry Grâce à*

Dieu and the *Mary Rose*, led the fleet into action. Then, with King Henry watching from Southsea Castle, the *Mary Rose* suddenly heeled over and sank. Only about 25 of the 400 men on board her survived. The rest were trapped by antiboarding netting that had been fixed over the upper decks. According to eyewitness accounts, the ship had her gunports open and her cannons run forward for action. As she turned toward the French, she heeled over and water flooded in through the open ports, turning the heel into a capsize. Meanwhile, the battle was inconclusive and the French withdrew.

There were several attempts to salvage the *Mary Rose* in the late 1540s, but they all failed. The ship had sunk so quickly that she had embedded herself in the muddy seabed, leaning over at an angle of

THE ANTHONY ROLL

The only contemporary illustration of the *Mary Rose* appears in a document called the Anthony Roll. In the 1540s, Anthony Anthony produced illustrations of 58 English naval vessels on three rolls of vellum as a gift for King Henry VIII.

They were later cut up and bound together in a book. The illustrations are not thought to be accurate in every detail, but they form a useful guide to the ships of the Tudor navy, including the *Mary Rose*.

about 60 degrees on her starboard side. Everything that was not buried was gradually consumed by marine organisms, and the *Mary Rose* disappeared. Occasionally, the shifting currents exposed the odd timber or two. It was one of these chance exposures in 1836 that enabled pioneer divers John and Charles Deane to spot the wreck. They recovered several guns, which revealed the wreck to be the *Mary Rose*, but it was soon forgotten again.

Rediscovery

Nothing more happened until 1965 when an amateur diver, Alexander McKee (1918–92), invited divers from the British Sub-Aqua Club's Southsea branch to begin mapping wrecks in the Solent. It was widely believed that wooden ships sunk hundreds of years earlier would have broken down and disappeared completely by then, but the divers found substantial parts of two 18th-century warships, the *Royal George* and the *Boyne*, still sitting on the seabed. McKee wondered if anything of the *Mary Rose* might have survived. In 1966, the divers found a depression in the seabed in a location where they thought the *Mary Rose* might be lying. The next year, a sonar scan of the seabed revealed a "buried anomaly" there.

In May 1971, divers finally located the wreck and began excavating the seabed to expose it. They tried to make sense of the timbers — their size, their angles, and so on — to figure out how much of the ship remained. Over the next few years, more than 600 volunteer divers

RAISING THE *VASA*

The only comparable project to raise an early modern warship was that of the Swedish warship *Vasa*. She was built in the 1620s, just over 100 years after the *Mary Rose*, and sank on her maiden voyage in 1628. Her sinking echoed that of the *Mary Rose*: She was pushed over so far by a gust of wind that water flooded in through her gunports. When attempts to raise her failed, she was abandoned and the location of the wreck was lost. She was rediscovered just outside Stockholm harbor in the 1950s and salvaged almost intact in 1961.

ABOVE: *Less than a year after being raised from the seabed, the remaining part of the* Mary Rose's *hull was put on display to the public. Millions of visitors have seen it, often under a misty spray of waxy preservative.*

worked to excavate the wreck, survey it and recover artifacts. Between 1979 and 1982, they made 28,000 dives. By 1982, the whole wreck had been uncovered. To the delight of the archaeologists, most of the starboard side of the ship was found to be intact, preserved from the destructive attentions of marine life by the airless mud that had covered it. Any timbers that could be removed were brought ashore, together with more than 19,000 smaller finds. This presented a huge challenge for conservators, who had to treat numerous wood and leather items to prevent them from breaking down and being lost forever.

If the wreck were to be left uncovered it would decay, so the decision was made to raise it. On June 15, 1982, a tubular steel frame was positioned over the remains, and wires were fixed between the frame and the hull. Then on September 28 a cradle, made in the shape of the hull and lined with air bags, was lowered onto the seabed. At the beginning of October, a floating crane lifted the frame, with the hull now hanging below it, and moved it over the cradle. The frame was lowered until the hull lay on the cradle's air bags. Then the frame, cradle and hull, weighing a total of 560 tons (570 metric tons), were lifted onto a barge and towed to the safety of the Royal Naval Base at Portsmouth. The operation was broadcast live on television and watched by 60 million people. It was extraordinary to see this Tudor warship emerging from the sea after more than 400 years.

ABOVE: *Vice-Admiral George Carew commanded the* Mary Rose *on the fateful day she sank. He had taken command earlier the same day.*

The hull was housed in a hall with chilled water spraying over it to keep it wet and prevent bacterial growth. About 800 timbers and deck planks that had been removed were reinstalled, and the water spray was replaced with an inert waxy chemical, polyethylene glycol (PEG). The wax gradually seeped into the wood and replaced the water. Then a more concentrated wax was used to seal the wood. In 2013 the sprays were switched off and the drying process began.

The Finds

Many of the pewter plates, tankards and spoons used by the ship's officers were found, including some stamped with the letters GC. These belonged to Vice-Admiral Sir George Carew (ca. 1504–45), the fleet's commander. The rest of the crew would have used wooden plates and drinking vessels. The ship's massive brick-built galley was found, with huge cooking pots and chopped wood stacked up nearby to fuel the fire. Dice, gaming boards, drums, pipes and leather book covers indicated how the sailors occupied their free time. There was a grindstone for sharpening knives and a tool called a heddle for weaving mats and repairing rigging.

The remains of about 200 men were found on board the ship. The leather hats, jerkins and shoes that they wore had been preserved well by the mud. Some woolen and silk items survived too. The crew of 200 mariners, 185 soldiers and 30 gunners included a contingent

LEFT: *The many personal possessions of the crew found in the wreck of the* Mary Rose *included a variety of combs.*

of archers. Although guns were replacing the bow and arrow by the 16th century, English longbowmen were still formidable in battle. One hundred and thirty-eight of their bows were found, together with more than 3,500 arrows.

In recent years, there have been several attempts to explain why the *Mary Rose* sank. The most popular theory is that the ship's refit, with the addition of more guns, had compromised its stability, making her heel over so much as she turned that water flooded in through her new gunports. However, the discovery of a French granite cannonball in the wrecked hull led some researchers to suggest that the ship may have been holed by a shot fired from one of the galleys that led the French attack. Water taken in through a hole in the hull would have sloshed to one side as the ship heeled over in a turn, making her lean more than normal, and then water entering through the gunports would have done the rest. We will never know precisely what happened, but if the *Mary Rose* had not sunk and ploughed into the muddy seabed, we would not have the ship and its contents to wonder at now.

VICTORIA

Some of the most audacious voyages of discovery were made as much for trade and conquest as for exploration. The first circumnavigation of the globe arose from the search for a sea route between Europe and the Spice Islands for the purpose of trade. The commander remembered as the leader of this historic expedition is the Portuguese sailor Ferdinand Magellan, but Magellan did not complete the voyage. Only one of his ships returned home after making it all the way around the world — *Victoria*.

TYPE: carrack

LAUNCHED: Gipuzkoa (Guipúzcoa), Spain, 1519

LENGTH: 59–69 ft (18–21 m)

TONNAGE: 85 tons

CONSTRUCTION: wood, carvel-planked

PROPULSION: square-rigged on the foremast and mainmast, lateen-rigged on the mizzenmast

After military service for Portugal in India, the Malay Peninsula and Morocco, Ferdinand Magellan (*ca.* 1480–1521) planned to search for a new sea-route from Portugal to the recently discovered Moluccas, or Spice Islands. When he failed to attract the interest or financial backing of the King of Portugal, he turned to Spain for help. The Spanish king, Charles I (later to become the Holy Roman Emperor Charles V), equipped Magellan with a fleet of five ships — *Victoria, Trinidad, San Antonio, Concepción* and *Santiago* — and enough provisions for 2 years. All the ships were carracks except for *Santiago*, which was a smaller caravel. Magellan chose *Trinidad* as his flagship. The crews of all five ships numbered about 270 men. A wealthy merchant contributed goods to trade. Magellan stood to gain a great deal from the voyage. If it was successful, the Spanish king agreed to grant him, among other things, a percentage of the proceeds of the voyage, an island of his own and the position of governor of any new lands claimed for Spain.

The fleet left Seville on August 10, 1519, and sailed down the Guadalquivir River to Sanlúcar de Barrameda. On September 20, the five

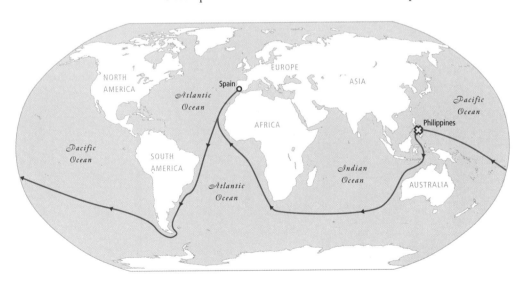

ABOVE: *Magellan's ships attempted to stay within sight of each other as they ventured into uncharted waters. Only one of the five ships,* Victoria, *would complete the circumnavigation of the globe and return to Spain.*

ships set sail across the Atlantic. The direction they took was determined by the Treaty of Tordesillas (1494), which divided all lands outside Europe between Portugal (lands to the east) and Spain (lands to the west). If Magellan had been sailing under the Portuguese flag, he could have sailed to the east, but as a result of the treaty he had to hope that the Spice Islands were in the western (Spanish) hemisphere, and so he sailed west. The Portuguese king was so outraged by Magellan's betrayal that he sent a naval force after him, but Magellan managed to avoid it.

Mutiny!

Magellan's fleet arrived in the bay of present-day Rio de Janeiro on December 13 and then followed the coast south. While it overwintered at Puerto San Julián in Patagonia, Argentina, at the end of March 1520, the crews of three of the ships mutinied. Magellan succeeded in putting down the revolt. The mutineers were executed or marooned on the coast and the expedition continued. One of the ships, *Santiago*, was wrecked in a storm while on a scouting mission farther down the coast.

The remainder of the fleet headed south on August 24. On October 21, they discovered the narrow channel through the tip of South America that led from the Atlantic to the Pacific. The channel was named the Strait of Magellan, after its discoverer. At this point, the *San Antonio* deserted from the fleet and returned to Spain. The three remaining ships — *Trinidad*, *Victoria* and *Concepción* — entered the Pacific Ocean on November 28. It was Magellan who named the ocean; he called it *Mar Pacífico* (Peaceful Sea). By March 1521 the fleet had crossed the ocean and reached the Philippine Islands, the first Europeans to do so.

LEFT: *Magellan's route took him across the Atlantic, down the coast of South America and around the southern tip of the continent, into the Pacific. Magellan was the first European to cross the Pacific.*

Local tribesmen guided them to Cebu, an island state where Magellan could provision his ships. He erected a cross and started baptizing the islanders, including their king and queen. On April 27, a rival tribe attacked them and Magellan was killed in what is known as the Battle of Mactan. The attackers carried off Magellan's body and refused to return it. A few days later, the two men elected to replace Magellan were also killed.

> THE CHURCH SAYS THE EARTH IS FLAT; BUT I HAVE SEEN ITS SHADOW ON THE MOON, AND I HAVE MORE CONFIDENCE EVEN IN A SHADOW THAN IN THE CHURCH.
>
> **Ferdinand Magellan**

The Return Journey

Now, with too few sailors to man all three ships, the survivors burned the *Concepción* and set sail in the two remaining ships, *Victoria* and *Trinidad*, under the command of João Lopes de Carvalho. They reached their destination, the Moluccas, on November 8, 1521.

By the time they were ready to begin the return voyage to Spain, laden with valuable spices, *Trinidad* was leaking so badly that she could not put to sea. Carvalho and his crew stayed behind with the ship, hoping to repair it and return later. Meanwhile *Victoria*, under the command of Juan Sebastián Elcano, crossed the Indian Ocean, rounded the Cape of Good Hope and finally reached Spain at the beginning of September 1522, completing the circumnavigation. *Trinidad* tried to

BELOW: *The first map of the Pacific Ocean, drawn by the Flemish cartographer Abraham Ortelius in 1589, shows Magellan's ship* Victoria *crossing the ocean.*

THE LOST DAY

When *Victoria* returned to Spain after circumnavigating the globe, her crew thought they had arrived on September 6. The actual date, however, was September 7. They had kept a careful daily log of their journey and couldn't understand how they could have lost a day. Nearly 60 years later, Sir Francis Drake experienced a similar lost day when he returned to England after his circumnavigation. Later, Dutch explorers reported the same strange phenomenon. Navigators eventually realized that a person traveling west loses 24 hours for every circuit of Earth. In the same way, someone traveling east gains 24 hours per circumnavigation. *Victoria*'s crew had discovered the need for an International Date Line, an imaginary line on the Earth's surface running from one pole to the other. Someone who travels west across the International Date Line adds one day to the date. Someone traveling in the opposite direction subtracts one day from the date. The line could be drawn anywhere on Earth, but it is most convenient to place it on the opposite side of the world from the prime meridian, the line of zero longitude. The prime meridian is drawn from pole to pole through Greenwich, England, so the International Date Line passes mostly through the middle of the Pacific Ocean.

return across the Pacific, but was captured by a Portuguese fleet. Only 18 of Magellan's sailors returned to Spain with *Victoria*. They had traveled 42,000 miles (68,000 km). Although the size of the world had been calculated in the past, *Victoria*'s voyage gave people a practical sense of the size of their home planet for the first time.

Spain concluded that the Moluccas were Spanish territory according to the Treaty of Tordesillas and sent a force including the *Victoria*'s commander, Elcano, to occupy the islands. Elcano was one of many sailors who starved to death during the voyage. Spain and Portugal finally settled their disagreement over the ownership of the Moluccas with the Treaty of Zaragoza in 1529, which assigned the Moluccas to Portugal and the Philippines to Spain. Magellan had blazed a trail for later explorers including Sir Francis Drake, who was to follow his route more than half a century later.

Victoria's historic importance went unrecognized. She was repaired and sold for use as a merchant ship, serving Spain's empire in the Americas. In about 1570, she sank with all hands in mid-Atlantic while sailing from the Antilles to Seville.

MAYFLOWER

In 1620 a group of Dissenters — Christians who had separated from the Church of England — traveled across the Atlantic to escape religious intolerance. Known as the Pilgrim Fathers, they founded a successful settlement in the land that would later become the United States of America. Although theirs was not the first European settlement in North America, it is the one that has woven itself into American folklore more than any other. These pioneers made their journey in a ship called the *Mayflower*.

TYPE: Dutch cargo *fluyt*

LAUNCHED: possibly Harwich, England, *ca.* 1580

LENGTH: about 100 ft (30 m)

TONNAGE: about 180 tons

CONSTRUCTION: wood, carvel planking

PROPULSION: sails on three masts and bowsprit

In the early 1600s in England, it was illegal not to attend Church of England services. Anyone who refused to attend was fined. Some Puritans (reformed Protestants) felt able to fall in line with the new laws, but others did not. Some of these Dissenters decided that they would have to leave the country. They moved to Amsterdam and then Leiden in Holland, where they had mixed fortunes. Some were able to find work, but for others the alien language and culture were more difficult to cope with. They were also uncomfortable with what they saw as the more liberal morals of the Dutch. Fearing the effects on their children, they decided to leave this country too. This time, they planned to travel to the New World where they could establish their own settlement and worship freely as they wished. They thought it better not to join the existing settlement in Jamestown, Virginia, where they might face the same disapproval and pressure to conform that they were trying to escape.

The younger and stronger members of the group would go first and then the older members would follow later. The first group departed from Delfshaven near Rotterdam in July 1620 in a ship called *Speedwell*. Originally named *Swiftsure*, she had taken part in the fleet that defeated the Spanish Armada in 1588. She was a 60-ton pinnace, a small square-rigged merchant ship. *Speedwell* sailed for Southampton, England, where she joined a second ship, the *Mayflower*, and more separatists who had decided to travel to the New World.

The *Mayflower* was a type of cargo ship called a *fluyt*. Originating in Holland, fluyts were designed to be inexpensive to manufacture, capable of carrying the maximum cargo on long voyages and easy for a small crew to sail. All of these factors cut the cost of transportation and so made them a popular type of cargo vessel in the 16th and 17th centuries. English shipbuilders were quick to see their advantages and started building their own fluyt-type ships. Although essentially cargo ships,

ABOVE: *The* Mayflower *was a typical 17th-century English merchant ship based on a Dutch design. Its high castles at the bow and stern gave the main deck some shelter from the elements.*

some of them were fitted out with cannons as warships. The date and place of the *Mayflower*'s construction are not known with any certainty, but the town of Harwich, Essex, claims to be her home. Some contemporary documents refer to the ship as the "Mayflower of Harwich," and her captain and part-owner, Christopher Jones, was from Harwich. Jones had been *Mayflower*'s captain for the previous 11 years, mainly carrying English wool to France and bringing back French wine.

Departure for the New World

Speedwell needed some repairs in preparation for the transatlantic voyage. The two ships finally left on August 5, 1620. *Speedwell* carried 30 passengers and *Mayflower* carried 90. *Speedwell* started taking on water almost immediately, so the ships put into Dartmouth, Devon, for repairs. When they left, *Speedwell* started leaking again and so both ships returned to Plymouth where *Speedwell* was abandoned. It was decided that *Mayflower* would make the voyage on her own with 102 passengers in addition to her crew of about 30. The passengers lived on the ship's gun deck, which measured only about 50 by 25 feet

LEFT: *Robert Walter Weir's painting of 1857,* Embarkation of the Pilgrims, *shows a group of the pilgrims gathered on deck, preparing for their voyage to the New World.*

ABOVE: *The pilgrims had very little space on board the* Mayflower. *Apart from the small main deck between the castles and the "tweendeck" or gun deck below it, the rest of the ship was full of cargo.*

(15 by 7.5 m) by just 5 feet (1.5 m) high. In fair weather they could use the main deck, but when storm clouds gathered, they were confined to the cramped gun deck which they shared with 10 cannons.

The *Mayflower* left Plymouth on September 6, 1620. Traditionally, Plymouth is regarded as the *Mayflower's* final port of call, although there is a claim that she visited Newlyn, Cornwall, to take on fresh water before setting out across the ocean. The ship was heading for the mouth of the Hudson River, but after a stormy passage lasting just over 2 months, she arrived at Cape Cod on November 11. The settlers explored and rejected several sites before they finally selected their new home. The area had already been named New Plymouth by the English explorer, Captain John Smith, who had surveyed it in 1614. The settlers kept the name, because of its fortuitous echo of their departure from England. The place where they came ashore is known as Plymouth Rock.

EARLIER EXPLORATIONS OF CAPE COD

The Pilgrim Fathers were not the first Europeans to visit the Cape Cod area. The Italian explorer John Cabot (ca. 1450–ca. 1500) was the first European to visit the North American mainland since the Vikings 500 years earlier (Columbus never set foot on North America). His discovery of Newfoundland in 1497 opened a new era of exploration along the east coast of the continent.

The French navigator Samuel de Champlain (1574–1635), who founded Quebec City, explored Cape Cod in 1605. John Smith (1580–1631), the English soldier and explorer who helped to establish the Jamestown colony, explored the Cape Cod area in 1614 and named it New England. His map of 1616 shows the site of the Plymouth Colony, which he called "New Plimouth."

Because of bad weather and illness, it took longer than expected to build the first houses, but by the end of January the settlers were able to start unloading the *Mayflower*. They encountered little opposition from native Americans, mainly because 90 percent of the indigenous population had been wiped out by an epidemic, probably smallpox.

Return to England

In April, the *Mayflower* set sail for England. Before she left, four of her cannons were unloaded to help defend the colony. After the arrival of more settlers with additional supplies in 1623, the surviving settlers gave thanks to God, an event that is still celebrated in the United States today as Thanksgiving. More settlers and the first cattle arrived the following year. By 1630, the colony had grown to 300 people and by 1691 the population was about 7,000.

Pushed along by strong westerlies, the *Mayflower* sailed home to England in half the time of the outward voyage. She arrived at Rotherhithe, London, on May 6, 1621. Her master, Christopher Jones, died the following year. The *Mayflower* lay at her berth for another 2 years. Nothing is known of her after that, but she was probably broken up. Some of the ship's timbers are said to have been used in the construction of a building known as the Mayflower Barn in Jordans, a village in Buckinghamshire, but this has never been confirmed.

BELOW: *This engraving by John C. McRae,* The Landing of the Pilgrim Fathers, in America, A.D. 1620 *(based on a painting by Charles Lucy), depicts the moment the pilgrims came ashore from the* Mayflower, *which appears in the background.*

HMS *Endeavour*

In the middle of the 18th century a humble coal ship took part in an expedition that changed our view of the world and its place in the solar system. At that time, the size of the solar system was unknown; no one knew how far away the Sun is. Observations made by scientists who voyaged into the Pacific Ocean aboard HMS *Endeavour* would resolve this mystery. The expedition's commander, James Cook, then headed for New Zealand and Australia on a secret mission.

TYPE: converted collier

LAUNCHED: Whitby, England, 1764

LENGTH: 98 ft (29.8 m)

TONNAGE: 366 tons

CONSTRUCTION: wood (white oak, elm and pine) carvel planking

PROPULSION: square-rigged sails on three masts and bowsprit

*E*ndeavour started her working life as a collier (coal ship) called the *Earl of Pembroke*. She was built in the coal port of Whitby on the northeast coast of England. Her hull was made of white oak, her keel and sternposts were elm and her masts were pine and fir. She was a strongly built ship with a flat bottom that enabled her to sail safely in shallow waters and also allowed her to be beached for repairs without needing a dry dock. In 1768, King George III approved plans for an expedition to the Pacific Ocean to observe the transit of Venus across the Sun the following year. The expedition had been requested by scientists who wanted to use the observation to help calculate the

RIGHT: *Sturdily built, flat-bottomed ships like Captain Cook's* Endeavour *were widely used in the coal trade on the northeast coast of England, where they were known as Whitby Cats.*

AMBITION LEADS ME NOT ONLY FARTHER THAN ANY OTHER MAN HAS BEEN BEFORE ME, BUT AS FAR AS I THINK IT POSSIBLE FOR MAN TO GO.
James Cook

distance between Earth and Sun. The Admiralty combined the expedition with a secret mission to look for a southern continent that was rumored to exist but had not yet been found or mapped.

James Cook (1728–79) was chosen to command the expedition. He had a reputation for highly accurate survey and mapping work. The Admiralty bought the *Earl of Pembroke* and had her converted from a collier into the scientific research vessel HMS *Endeavour*. The conversion included the addition of an extra deck of cabins and storerooms. The crew of 85 included a dozen Royal Marines. There was also a naturalist (Joseph Banks), an astronomer and two artists on board. The ship might have to defend herself against attack, so she was armed with six cannons and 12 swivel guns.

Endeavour left Plymouth on August 26, 1768. Cook had to reach Tahiti, a tiny island less than 30 miles (45 km) across in the middle of the vast Pacific Ocean, before the following June. He reached Cape Horn in the middle of January. After 3 days battling against storms and tides, he finally sailed into the Pacific. He stopped for Banks to collect plant samples from the coast and then set out across the ocean for Tahiti, which he reached in April.

The crew set about building an observatory on the shore. It was surrounded by earth banks topped with wooden palisades and defended by guns from the ship. When the transit occurred on June 3, it was difficult for the observers to time it precisely because of a blurring of the edge of the planet Venus as it neared the edge of the Sun's disk; this is called the black drop effect. Nevertheless, using observations from Tahiti and elsewhere, the average distance between Earth and Sun was calculated to be 93,726,900 miles — an error of only 0.8 percent, as the correct distance is 92,955,807 miles (149,597,870 km).

ABOVE: *Cook's official
portrait was painted by
Nathaniel Dance-Holland.
It shows the great navigator
in his Royal Navy full-dress
captain's uniform, his hand
resting on his own chart
of the Southern Ocean.
He is pointing at the
coast of Australia.*

Terra Australis

With the transit completed, Cook opened sealed orders that instructed him to search for the southern continent, or *Terra Australis*. He sailed for New Zealand and mapped its entire coastline, thus proving that it wasn't part of a huge continent to the south. Then he continued west and reached the east coast of Australia, the first European to do so. This, too, was not part of a bigger continent thought to be lying to the south. He claimed the land for Britain and called it New South Wales. On April 29, he made landfall at a place he named Botany Bay. As she was leaving, *Endeavour* ran aground on the Great Barrier Reef and her hull was holed. It was patched temporarily with a piece of sail. When the crew found a suitable place for repairs, they beached the ship. Then they continued to Batavia (Jakarta) in the Dutch East Indies for supplies and more permanent repairs to the leaking hull, which proved to be in far worse condition than anyone had thought. Many of the crew succumbed to malaria and dysentery, which were raging through the Dutch colony like a forest fire. Cook continued west with his surviving crew, completing his first circumnavigation in July 1771, 3 years after he had left.

Cook would go on to make two more voyages of discovery, but not in *Endeavour*. For these voyages, Cook chose another converted collier, HMS *Resolution*. During his second circumnavigation (1772–75), he

SCURVY

James Cook had spent more than 20 years at sea before commanding the 1769 expedition in *Endeavour*, so he was very well aware of the lives led by ordinary seamen. After he became an officer, he took good care of the men who served under him. He was particularly concerned about the food his crews ate. He ensured that their diet included pickled cabbage and preserves of oranges and lemons. He ordered all sailors to eat the food provided; any who refused were flogged. Cook's insistence paid off. He was the first British naval commander to eliminate scurvy, an illness caused by a deficiency of vitamin C, from his crews.

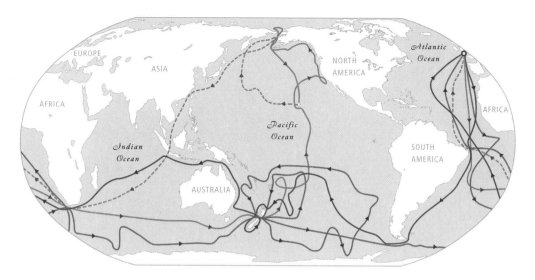

ABOVE: *Cook's three voyages of exploration yielded an enormous amount of information about the little-known Pacific. Here his first voyage is shown in red, his second is in green and his third is in blue; the dashed blue line shows the route his crew took after his death.*

looked for the southern continent again. He sailed far enough south to reach the pack ice and even circumnavigated Antarctica, but he never sighted land. He thought the pack ice might extend all the way to the South Pole.

The Northwest Passage

In 1776, Cook left on an expedition to search the north Pacific coast of America for the northwest passage linking the Atlantic and Pacific oceans. On the way, he discovered the Hawaiian islands. After he encountered pack ice in the Bering Strait, he returned to Hawaii. While he was there, one of his boats was stolen. He tried to take the local chief hostage for the boat's return, but in the ensuing scuffle Cook was attacked and killed. Four marines died too. It was February 14, 1779. The native people took Cook's body away with them, though some of his remains were later returned for burial at sea.

Meanwhile, HMS *Endeavour* served as a naval transport ship until she was sold to a private buyer in 1775. Soon afterward she was declared unseaworthy and needed extensive work before being used again as a troop transport and then a prison ship. Now called the *Lord Sandwich*, she was scuttled in 1778 along with a number of other naval and private ships to blockade Narragansett Bay in Rhode Island to protect the British settlement there from a French naval attack.

HMS *Victory*

The Battle of Trafalgar in 1805 is one of the most famous naval battles in British history and the British fleet's admiral, Horatio Nelson, is one of the most famous British military figures. Nelson's flagship at Trafalgar was HMS *Victory*. Trafalgar finally saw off the French threat of invasion and established Britain as the world's predominant sea power. *Victory* is still a commissioned warship in the Royal Navy, making her the world's oldest naval ship still in commission.

TYPE: first-rate ship
of the line

LAUNCHED: Chatham
Dockyard, Kent, England,
1765

LENGTH: 227 ft 6 in (69.3 m)

TONNAGE: 3,500 long tons
(3,556 metric tons)
displacement

CONSTRUCTION: oak hull,
carvel planking

PROPULSION: square sails on
three masts and bowsprit

*H*MS *Victory* was ordered for the navy during the Seven Years' War, a conflict between an alliance of nations led by Britain and another led by France. She was built at Chatham Dockyard on the River Medway in Kent. Her keel was laid on July 23, 1759. A forest of around 6,000 trees, mostly oak, was used in her construction. By the time she was launched in 1765, the war was over and so she remained unfinished "in ordinary" (held in reserve). She was finally completed in 1778 after the American Revolutionary War had broken out. Armed with 104 guns, making her a "first-rate" warship, she saw action against the French at the First and Second Battles of Ushant. In 1782 she took part in the Battle of Spartel against the combined fleets of France and Spain during the relief of Gibraltar. Then in 1797 she served as the flagship of Admiral Sir John Jervis at the Battle of Cape St. Vincent against the Spanish.

RIGHT: Victory *was quite an old ship, in service for 40 years, when Nelson paced her quarterdeck at the Battle of Trafalgar in 1805.*

FIRST-RATE WARSHIPS

British warships at the time of HMS *Victory* were divided into six groups called "rates" according to their size and the number of guns they carried. First-rates like *Victory* were the biggest and most powerful; each was armed with at least 100 guns on three decks and carried a crew of more than 800 officers and men. They often served as an admiral's headquarters, or flagship — so called because a special flag was flown to signify his presence.

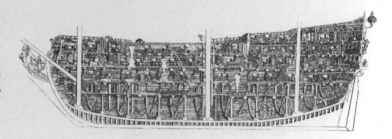

RIGHT: Victory *was an example of the supreme naval fighting machine of her day — a first-rate British Royal Navy warship. This illustration from* Cyclopaedia *(1728) compares a third-rate warship (top) with a first-rate like* Victory.

By then she was in very poor condition. She was refitted as a hospital ship and then almost became a prison hulk, the last stop before the breaker's yard. However, when another warship, HMS *Impregnable*, was wrecked in 1799, *Victory* was brought back into service and updated according to the latest naval standards. Her hull was covered with 3,923 copper sheets to protect the wood from shipworm. Her heaviest guns, 42-pounders, were replaced with lighter 32-pounders that could be loaded and fired faster. It was at this time that her hull was painted in the famous black and yellow stripes.

Trafalgar

Victory's finest hour came in 1805. French emperor Napoleon Bonaparte had abandoned his ambition to invade Britain, but his fleet of French and Spanish ships still represented a threat. The British decided to deal with the threat and Vice-Admiral Lord Nelson (1758–1805) was given the task. He joined the British fleet near the Spanish port of Cádiz, where the enemy ships lay. When the French and Spanish ships set sail, the British followed them out into the open sea and attacked them on October 21 off Cape Trafalgar. Before the battle, Nelson sent his famous signal to the fleet: "England expects that every man will do his duty."

Nelson knew he was outnumbered — 27 British ships versus 33 French and Spanish — so he decided to employ a risky tactic called "breaking the line." At that time, warships usually fought in straight lines, with the two lines alongside each other, blasting one another to bits with their cannons. Nelson chose a different approach. He divided his ships into two groups. One headed straight for the middle of the enemy line of ships and broke through them, while the other British group broke through at the rear of the line, dividing the enemy formation into smaller groups. As the British ships approached, they were on the receiving end of enemy gunfire for almost an hour before their own guns could be brought to bear on their targets. Nelson had gambled that the enemy gunners would be ineffective because they had seen very little action, whereas the British gunners were fresh from victories at the Nile and Copenhagen. He was right. British naval gunners at that time were so experienced and well-drilled that they could fire every 90 seconds — at least twice the firing rate of any other gunners. Once the British ships had broken the enemy line, they could finally attack individual ships. *Victory* knocked out the French Admiral's flagship, *Bucentaure*, first. Nelson's gamble and superior British gunnery won the day. Nineteen enemy ships were captured or destroyed on the day of the battle. Another four ships were captured later and most of the others were reduced to floating wrecks. No British ships were lost.

BELOW: *The most famous painting of Nelson was made by Lemuel Francis Abbott in 1800. Nelson's empty right sleeve is pinned across his chest, a reminder that he had lost his arm at Santa Cruz in 1797.*

ENGLAND EXPECTS THAT EVERY MAN WILL DO HIS DUTY.
Horatio Nelson's signal to the British fleet before the Battle of Trafalgar

HMS *VICTORY* AT TRAFALGAR

At the time of the Battle of Trafalgar in 1805, HMS *Victory* carried a crew of 821. More than 500 of them were the seamen who sailed the ship. About 289 of these were volunteers. The rest, more than 200 men, had been pressed into service; when the navy needed men and there were not enough volunteers, press-gangs had the power to take suitable men into service by force. In addition, *Victory* had about 150 marines. The rest of the crew were the ship's officers and specialists, including the surgeon. About 40 percent of Victory's crew were under the age of 24. The youngest member of the crew was only 12 — the same age as Nelson when he joined the navy — and the oldest was the 67-year-old purser.

ABOVE: *Nicholas Pocock's painting was made in 1807, only 2 years after the battle. It shows the situation at 5 p.m. on October 21, 1805, as the battle is drawing to a close.*

Nelson's Death

Sailors on a warship's upper decks risked being shot by enemy gunners and snipers. The poop deck and quarterdeck were particular targets because the ship's senior officers were concentrated here. *Victory's* wheel on the quarterdeck was shot away and the ship had to be steered from her tiller below decks. Nelson was standing on the quarterdeck when he was hit by a musket ball fired by a sniper from the rigging of the French warship, *Redoutable*. The two ships were just a few feet apart. The ball entered Nelson's right shoulder and passed through his spine. When *Victory's* captain, Thomas Hardy, saw that Nelson had been hit, he ran to help him. Nelson was carried down to the orlop deck (the lowest deck). He was laid in the cockpit, the part of the deck where wounded men were brought for the attention of the ship's surgeon. Nelson told the surgeon, "You can do nothing for me. I have but a short time to live. My back is shot through." When Hardy came down to see him, Nelson uttered the famous words, "Kiss me, Hardy." The surgeon saw Hardy kiss Nelson on the cheek and forehead. Nelson faded and died at 4.30 p.m., 3 hours after he had been shot.

Victory had taken a heavy pounding during the battle and was further battered by a storm. Despite the severe damage the ship had suffered, the crew demanded the right to bring their commander home. Nelson's body was preserved in a cask of brandy. The ship was so badly damaged that she had to be towed to Gibraltar for repairs before returning to England under tow. Nelson was given a state funeral and his body was laid to rest in a tomb below the dome of St. Paul's Cathedral in London.

NELSON'S CAREER

Nelson was born on September 29, 1758, in Burnham Thorpe, Norfolk, England. He joined the navy at the age of only 12. Despite suffering from seasickness, he rose rapidly through the ranks, becoming a captain at 20. After service in the West Indies, he returned to England and spent 5 years waiting for a command until he was finally made captain of HMS *Agamemnon* in 1793. After losing the sight in his right eye at Corsica, he was given command of HMS *Captain*. He played an important part in the Battle of Cape St. Vincent (1797), which earned him a knighthood and a promotion to Rear Admiral. Later the same year, he lost his right arm at the Battle of Santa Cruz. In 1798, a British fleet under his command defeated the French at the Battle of the Nile.

Nelson had a habit of ignoring orders if he felt he could do better. When ordered to withdraw at the Battle of Copenhagen (1801), he famously put his telescope to his blind eye and claimed he could not see the signal. After victory at Copenhagen, he became a viscount and was promoted to commander-in-chief.

RIGHT: Victory's gun decks are calm, quiet, neat and tidy today, but in the heat of battle on a rolling sea, they were places of great military professionalism and also terrifying noise, smoke, smell and violence.

BELOW: HMS Victory before being restored to the slightly paler colors she wore at the time of the Battle of Trafalgar. The ship continues to attract visitors at the Portsmouth Historic Dockyard on the south coast of England.

Victory **Today**

Victory was given a major refit. She then saw service in the Peninsular War and in the Baltic before being taken out of service in 1812 and moored at Portsmouth. In the 1920s naval historians warned that if she were left to rot any longer, she might soon be beyond rescue. Some restoration work was done, but she was badly damaged by bombing during World War II. By the 1970s, she needed urgent repairs. During her service life, she had gone through several changes of armament, rigging and outward appearance. The decision was made to restore her to her 1805 Trafalgar condition. Today, only about 20 percent of the original ship remains, but the lower gun deck and the orlop deck where Nelson died are mostly original.

Research and restoration have continued to the present day. Most recently, researchers discovered the precise colors of *Victory*'s paint scheme at the Battle of Trafalgar: not black and bright yellow, as had been thought, but graphite grey and a pale creamy-orange yellow. A 4-month repainting project was completed in October 2015.

HMS SIRIUS

The arrival of the First Fleet in Botany Bay in 1788 marked the beginning of the European settlement of Australia. The fleet's flagship was a converted merchantman, HMS *Sirius*. Her senior officers included the first three governors of New South Wales.

TYPE: merchantman converted to a 10-gun naval ship

LAUNCHED: Rotherhithe, England, 1780

LENGTH: 110 ft 5 in (33.7 m)

TONNAGE: 512 tons

CONSTRUCTION: wood, carvel planking

PROPULSION: square sails on three masts

England had been transporting some of its convicted criminals to the Americas since the early 17th century, but the loss of the American colonies at the end of the American Revolution closed that route. British prisons quickly became even more overcrowded than usual. Old ships known as "hulks" were pressed into service as floating prisons. When these were full, a new overseas destination for penal transportation was urgently needed. Just a few years earlier, Captain James Cook had discovered a whole new continent, Australia. In 1785, the British government decided to transport criminals to this new land on the other side of the world.

The First Fleet, as it became known, left Portsmouth to establish the first British penal colony in Australia on May 13, 1787. It consisted of 11 ships — two Royal Navy vessels, three stores ships and six convict ships — under the overall command of Admiral Arthur Phillip (1738–1814) in his flagship, HMS *Sirius*, captained by John Hunter. Phillip was also appointed Governor-designate of New South Wales, the land claimed for Britain by James Cook. Sources vary as to the precise number of convicts transported by the fleet, but there were about 750 men, women and children. Most of them had been found guilty of robbery or theft.

RIGHT: *The First Fleet arrives off the coast of Botany Bay in an illustration by engraver Thomas Medland of a painting by Robert Clevely. The illustration was produced in 1789, just 1 year after the event.*

From *Berwick* to *Sirius*

Sirius had started out as a merchant ship called *Berwick*. After being damaged in a fire, she was bought by the Royal Navy and refitted as a military vessel in Deptford Dockyard. The refit included arming her with 10 guns and sheathing her hull with copper to protect the wood from marine life. She served as HMS *Berwick* in American waters from 1782 until the end of the Revolutionary War, and was then stationed in the West Indies until she was paid off (removed from active service and the crew discharged) at the beginning of 1785. The following year, she was selected for the First Fleet and prepared for service again. On October 12, 1786, she was renamed HMS *Sirius* after the Dog Star, the bright star in the southern constellation Canis Major.

After a supply stop in the Canary Islands at the beginning of June, the ships reached Rio de Janeiro at the beginning of August. A month later, they sailed for the Cape of Good Hope, their last stop before Australia. Conditions for the convicts were very unpleasant and insanitary. They were kept below decks for much of the voyage. Their clothes were infested with lice and fleas. Several convicts fell ill and died.

The ships began arriving at Botany Bay on January 18, 1788, after a 252-day, 15,000-mile (24,000-km) voyage. While the settlers were exploring the bay, two French ships arrived. They were making an around-the-world voyage to check and complete the maps made by Captain Cook and establish new trading links in the Pacific. As the British ships planned to return to Europe first, the French gave them some documents, letters and charts to take back with them.

> THERE ARE FEW THINGS MORE PLEASING THAN THE CONTEMPLATION OF ORDER AND USEFUL ARRANGEMENT.
>
> **Sir Arthur Phillip**

The French ships departed on March 10 and were never seen again. Interestingly, one of the men who had applied to join this French expedition was a young Corsican called Napoleon Bonaparte. If he had been accepted and had perished with the rest of the seamen, European history would have been very different.

Meanwhile, Botany Bay was judged to be unsuitable for a penal colony. It was difficult to defend and the water was too shallow for the ships to anchor close to shore. When the French ships left, *Sirius* helped to move everyone north to a better location. They chose Port Jackson, which had been named by Captain Cook. They anchored in a cove named Sydney by Arthur Phillip after the British Home Secretary, Lord Sydney. The site of Sydney Opera House was once called Cattle Point, because it was used to hold cattle and horses brought by the First Fleet. The cove had a supply of fresh water and the soil was better there than at Botany Bay.

With the settlers safely ashore and the first houses built, *Sirius* left on October 2, 1788, to collect more supplies from the Cape of Good Hope. By the time she returned, more than 7 months later, the colonists were close to starvation. A ship sent from England with more supplies had failed to arrive, and the situation was critical. To relieve the pressure on the Port Jackson settlement, some of the convicts were sent to Norfolk Island, where a small settlement had been established soon after the fleet arrived in Australia. On March 19, 1790, while *Sirius* was landing convicts and marines on the island, she was blown onto a reef and sank. The crew managed

BELOW: *HMS* Sirius *comes to grief as she founders on a reef off the coast of Norfolk Island on March 19, 1790.*

THE LONGITUDE PROBLEM

Every point on the Earth's surface is identified by two numbers: latitude (degrees north or south of the equator) and longitude (degrees east or west of the prime meridian). Sailors determined their latitude by measuring the height of the Sun above the horizon at noon, but longitude was more difficult to measure. It required a sailor to know the difference between the local time and the time at the prime meridian. And that required a very accurate clock that would keep good time for weeks or months at sea on a pitching, rolling ship. HMS *Sirius* is thought to have carried a very accurate timepiece called the K1 marine chronometer, which had also been carried by James Cook on his second and third voyages. The K1 was a copy of a chronometer made by John Harrison (1693–1776). Harrison had won a British government competition to make an accurate marine chronometer. He spent 40 years building a series of increasingly accurate timepieces that finally cracked the problem of determining longitude at sea.

to reach the island, where they remained stranded until they were rescued a year later and returned to England. Moving some of the convicts to Norfolk Island proved to have been wise, because it bought more time for Port Jackson to survive until supply ships arrived. The remains of the *Sirius* at Norfolk Island are the only in situ remains of a First Fleet ship.

Two more convict fleets arrived in 1790 and 1791, with the first free settlers arriving in 1793. New South Wales remained a penal colony until 1823. By the time the last convict ship arrived in Western Australia in 1868, 162,000 convicts had been transported from Britain in 806 ships. The total population of all the Australian colonies at this time had grown to about one million.

RIGHT: *Larcum Kendall's K1 marine chronometer was carried on board HMS* Sirius *and enabled the crew to establish the ship's position with great precision. It was a copy of John Harrison's H4 chronometer.*

CLERMONT (NORTH RIVER STEAMBOAT)

Robert Fulton's *Clermont* was the world's first commercially successful steamboat. It wasn't the first steamboat but it was the one that proved steam power to be a practical and commercially successful propulsion technology for ships and boats. In doing so, it changed the world of marine transport by freeing ships from dependence on the wind.

TYPE: steamboat

LAUNCHED: New York, 1807

LENGTH: 142 ft (43 m)

TONNAGE: 121 tons displacement

CONSTRUCTION: wood, carvel planking

PROPULSION: steam-powered paddlewheels and sail

RIGHT: *Robert Fulton, a painter-turned-inventor, created the first commercially successful steamboat and practical submarines.*

Robert Fulton was born on a farm in the township of Little Britain, Pennsylvania, in 1765. He became interested in steam engines in his childhood, but his first love was art. He spent 6 years working as an artist in Philadelphia and did so well that he was able to buy a farm for his mother.

He traveled to England in 1786 and spent the next 10 years working as an artist, but his interest in machines continued. He built an experimental steamboat for the 3rd Duke of Bridgewater (1736–1803) for use on the duke's canal, but the project was abandoned over concerns that the boat's paddles might damage the canal's clay lining. In 1797 Fulton moved to Paris, France, where, 14 years earlier, the Marquis de Jouffroy d'Abbans (1751–1832) had managed to get a steamboat to work on the Saône River for all of 15 minutes. John Fitch (1743–98) built a more successful steamboat, which he trialed on the Delaware River in 1787.

A Successful Partnership

While Fulton was in France, he designed and built the first practical submarine, *Nautilus*. Then he met Robert R. Livingston (1746–1813), the U.S. Ambassador to France. Like Fulton, Livingston had experimented with steamboats. The two men teamed up and decided to build a steamboat capable of carrying passengers up and down the Hudson River. The first boats they built were disappointing. Fulton moved back to England, where he invented a torpedo for the Royal Navy and started work on a second submarine. However, after Nelson's victory at the Battle of Trafalgar, the French threat receded and Fulton's services were no longer required. He returned to the United States in 1806 and married Livingston's niece.

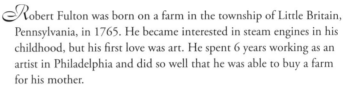

ABOVE: *A full-size replica of the* Clermont *was built in 1909. After serving as a museum for several years, she was left to decay until she was broken up for scrap in 1936.*

CHARLOTTE DUNDAS

The first practical steamboat was the *Charlotte Dundas*, built by William Symington (1764–1831). The 56-foot (17-m) vessel was built in 1802 as a tug to tow barges on the Forth and Clyde Canal in Scotland. Powered by a 10-horsepower (7.5 kW) steam engine driving a single paddlewheel at the stern, it proved itself by pulling two 70-ton barges a distance of 19.5 miles (31 km) in 6 hours. It was, however, a commercial failure, because potential customers feared that the wash from its paddlewheel might damage the canal banks.

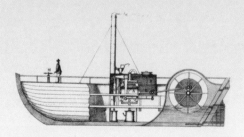

Fulton had a British Boulton and Watt steam engine shipped out to the United States and Livingston secured a monopoly to operate steamboat services on the Hudson. Using the British engine, one of the most advanced steam engines of its time, they built the steamboat that made history. It was a wooden boat 142 feet (43 m) long with a beam (width) of 18 feet (4.3 m). The steam engine was mounted amidships turning two paddlewheels, one on each side of the boat. The paddlewheels were 15 feet (4.6 m) in diameter and 4 feet (1.2 m) wide. A pile of pine logs fueled the engine.

The boat is known as the *Clermont* today, but it was never known by that name in Fulton's time. It was mistakenly called *Clermont* — the name of Livingston's home on the Hudson River — in a book about Fulton published in 1817, and the name stuck. When it began carrying paying passengers, it was advertised as the *North River Steamboat*. Passersby who saw it being built were convinced it would fail, so much so that it was nicknamed "Fulton's folly."

In 1800, Robert Fulton designed the world's first practical submarine, *Nautilus*. The 21-foot (6.5-m) long vessel was made of copper sheets formed around iron frames. When on the surface, it was propelled by a sail. When it submerged, by letting water into its hollow keel, the crew turned a hand-cranked propeller. *Nautilus* was designed as a military vessel to attach a mine to an enemy ship. It was tested successfully in the Seine River in France in 1800 and then in the sea at Le Havre. Despite this, the French navy decided that it was far too dangerous to its own crew to be a useful naval vessel, and it was dismantled.

LEFT: *This replica at Cité de La Mer, in Cherbourg, France, shows Fulton's Nautilus with its sail raised.*

The Maiden Voyage

The new boat was launched on August 17, 1807, and tested for the first time. Fulton was irritated to discover that Livingston had turned the experimental try-out into a public demonstration with invited guests. The guests' doubts about the boat seemed to be confirmed when its engine suddenly stopped just after they set off. Fulton made a quick adjustment and restarted the engine, which gave no further trouble. The boat chugged its way 150 miles (240 km) up the Hudson River from Manhattan to Albany in 32 hours, including an overnight stop. Sailboats took up to 6 days to make the same journey. By all accounts the engine was very noisy. It was described as "a backwoods saw-mill mounted on a scow (flat-bottomed barge) and set on fire." When the boat stopped at Livingston's home, the engine was so loud that Livingston couldn't make himself heard above the noise. As it continued on its way, crowds gathered on the banks to see it pass. However, so many of the passengers feared that the flaming, smoking engine might explode that only two of them stayed on board for the return journey to New York.

Following the successful trial, the boat was prepared for passenger services. In only 2 weeks passenger cabins were added, an engine cover

RIGHT: *Fulton's interest in submarines continued after* Nautilus. *He experimented with further designs, including this proposal dating from 1806.*

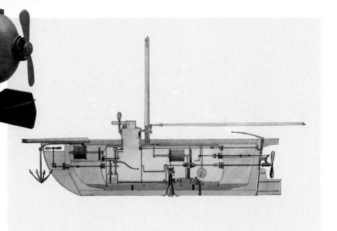

ABOVE: *This view from the deck of the 1909 replica of the* North River Steamboat *shows the paddlewheel on each side of the hull, driven by the steam engine in the middle of the vessel.*

was fitted and guards were built over the paddlewheels to reduce the spray they threw up. On September 4, 1807, regular scheduled passenger services began. The boat made the trip to Albany and back every 4 days with up to 100 passengers, until the winter when floating ice made the river too dangerous. As news of the boat's success spread, Fulton was asked to build similar boats for operators up and down the East Coast.

The *North River Steamboat* was retired in 1814 and its fate is unknown. Fulton went on to design the *Demologos*, the world's first steam-powered warship, for the U.S. Navy. It was not completed until 1815, after Fulton's death at the age of 49. He died after being drenched with icy water and developing pneumonia, followed by tuberculosis, when he rescued a friend who had fallen through the ice while walking on the frozen Hudson River. The warship's name was changed to *Fulton* in his honor.

MY STEAMBOAT VOYAGE TO ALBANY AND BACK HAS TURNED OUT RATHER MORE FAVORABLY THAN I HAD CALCULATED.
Robert Fulton (referring to his steamboat's maiden voyage in 1807)

SS SAVANNAH

The numbers of steamboats plying their trade on rivers and coasts mushroomed in the early 1800s. Their great advantage over sail was that they could go where they wished at the time of their choosing, without having to rely on the wind blowing hard enough in the right direction. They could stick to a schedule. But no steam-powered ship had ever crossed the Atlantic Ocean and some people thought it might never happen — until one American sea captain dared to dream.

TYPE: hybrid sailing ship and sidewheel steamer

LAUNCHED: New York, 1818

LENGTH: 109 ft (33 m)

TONNAGE: 320 tons

CONSTRUCTION: wood, carvel planking

PROPULSION: full-rigged sails on three masts and bowsprit plus two steam-powered paddlewheels

The *Savannah* was designed as a sailing packet — a ship for carrying mail — but even before she was launched, she was converted to steam power for a historic voyage. A prominent Connecticut sea captain, Moses Rogers, persuaded a consortium of wealthy backers in Savannah, Georgia, to buy the ship, install a steam engine and attempt the first steam-powered Atlantic crossing. Savannah was one of the most important American ports in the early 18th century, and Rogers was keen that it should achieve this historic first. It could inaugurate a lucrative transatlantic cargo and passenger service. The ship would be named after its home port.

Rogers had seen the maiden voyage of Fulton's *North River Steamboat* in 1807. Two years later, Rogers was the captain of the *Phoenix*, the first practical all-American steamship. When he sailed the *Phoenix* down the East Coast from New York to Philadelphia, it was the first ever ocean voyage by a steamship. By 1817, he was captain of the steamship *Charleston* on its passenger service between Charleston and Savannah.

ABOVE: *The Savannah River was a busy transport route between the ocean and the interior in the 19th century. The SS* Savannah *was intended to extend trade potential across the Atlantic.*

Converting the Ship

The conversion work on the *Savannah* was extensive. The steam engine was fitted amidships driving two 16-foot (4.9-m) wrought-iron sidewheels. The wheels were designed to be collapsible so that they could be removed and stored on deck when the ship was under sail. The engine vented smoke through a curious 17-foot (5.2-m) bent smokestack that could be turned according to the wind direction.

The ship's passenger accommodation was expensively decorated and furnished with mahogany paneling and imported carpets. There were 16 two-berth staterooms, with separate men's and women's quarters, and three saloons. Mirrors were widely used to create an illusion of extra space.

Doubts about the performance and safety of steam power at sea made it difficult to recruit a crew for the passage. After failing to hire a crew in New York, where the ship was built, Rogers had more success in his hometown of New London, Connecticut. When the construction and conversion work was complete, the ship was moved to Savannah.

On May 11, 1819, President James Monroe visited Savannah and accepted an offer of a short excursion on the ship. He was so impressed that he asked for her to be brought to Washington so that Congress could see her. He thought she might be an effective antipiracy vessel for the Caribbean, so the government might be prepared to buy her. But, for the time being at least, her owners had other plans for her.

Crossing the Ocean

Efforts to attract passengers for the transatlantic voyage failed. The chances of sinking in mid-ocean were thought to be so high that the ship was nicknamed the "steam coffin," so at 5 a.m. on May 24 *Savannah* left port and headed out into the ocean with only her crew on board. Her captain was Moses Rogers. She carried 75 short tons (68 metric tons) of coal and 25 cords of wood. The cord is an old unit of volume. A cord of wood is a pile measuring roughly 4 feet high by 8 feet long and 4 feet across (1.2 by 2.4 by 1.2 m).

Savannah started under both sail and steam power. During the voyage, the crews of at least two other ships saw the thick, black smoke rising from her smokestack and, as a steamship had never been seen in mid-ocean before, they thought she must be a sailing ship on fire, but

LEFT: *In common with many early steamships, the SS* Savannah *had both sails and steam propulsion. Sails were not dispensed with until crews and passengers gained more confidence in the safety and reliability of steam propulsion.*

could not catch her to render assistance! She ran out of fuel as she passed the Irish coast at Cork and reached Liverpool, England, at 6 p.m. on June 20. She had used her steam engine for a total of 80 hours during the crossing.

She was a huge attraction in Liverpool, drawing thousands of sightseers. Nearly a month later, she departed for Russia via Denmark and Sweden. In Sweden, she welcomed her first passengers aboard and Rogers turned down an offer from the Swedish government to buy her. When she docked in St. Petersburg, she attracted as many curious visitors as in Liverpool. The Tsar offered Rogers the sole steamship rights for all Russian waters. It was a handsome offer, but Rogers declined because he was keen to get back home. *Savannah* left Russia on September 29 and after stops in Denmark and Norway she set out on the return journey across the ocean, arriving at Savannah on November 30.

Despite the success of the historic voyage, she was unable to attract enough passengers or cargo to establish a regular transatlantic service. Her owners tried to take up President Monroe's offer to buy her, but the government had lost interest

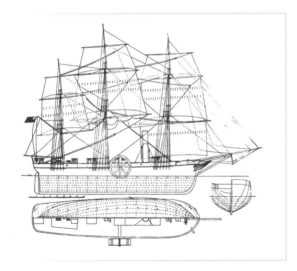

LEFT: *Drawings show side-mounted paddlewheels that seem too small for the* Savannah, *but they were intended for use only when there was too little wind for the sails.*

by then. A serious fire destroyed most of downtown Savannah on January 1820, causing further financial problems for the ship's owners. They converted the *Savannah* back to sail only. Without a steam engine, she could carry a lot more cargo. She served as a sailing packet (mail ship) between New York and Savannah until she ran aground at Long Island on November 5, 1821, and broke apart.

Ships like *Savannah* were too small to carry enough fuel for a whole transatlantic voyage under steam power, and the engines were so big and inefficient that they needed a great deal of fuel. But bigger ships with more efficient engines would soon be built, heralding a new age of transatlantic sea travel.

WHO WAS FIRST?

History records the SS *Savannah* as the first steamship to cross the Atlantic, in 1819, but there are other claimants. Their claims are based on the fact that the *Savannah* was under steam power for only a small fraction of her passage, so she was really a sailing ship with an auxiliary steam engine, not a steamship. One alternative claimant is the British-built, Dutch-owned vessel, *Curaçao*, which crossed from Holland to Curaçao in

the Caribbean in 1827. Another is the Canadian steamship *Royal William*. When her owners decided to sell her in 1833, they sailed her across the Atlantic from Nova Scotia to England to look for a buyer.

BELOW: *The paddle steamer* Curaçao *began a regular transatlantic service in the 1820s, but she made only three crossings before the service was abandoned.*

HMS *BEAGLE*

When an unknown naturalist was offered a spare place on a British survey and mapping voyage to South America, no one could have imagined the global consequences that would result. The naturalist was Charles Darwin and the ship was HMS *Beagle*. Observations made by Darwin during the voyage led to his revolutionary theory of evolution by natural selection.

TYPE: *Cherokee*-class brig-sloop converted to a bark

LAUNCHED: Woolwich Royal Dockyard, England, 1820

LENGTH: 90.3 ft (27.5 m)

TONNAGE: 235 tons (242 for second voyage)

CONSTRUCTION: wood, carvel planking

PROPULSION: square sails on three masts and bowsprit

ABOVE: *Mount Sarmiento towers over HMS* Beagle *as the famous survey vessel with Charles Darwin on board enters the Straits of Magellan in 1834. This image illustrated the frontispiece of Darwin's* Journal of Researches *(1839).*

ℋMS *Beagle* was built as a type of small two-masted warship called a brig-sloop. There were two classes of brig-sloop: the 18-gun *Cruizer* class and the 10-gun *Cherokee* class. *Beagle* was a 10-gun *Cherokee*-class brig-sloop. More than 100 of these were built for the Royal Navy. Unfortunately, so many of them were lost at sea that they became known as "coffin brigs." Of 107 built, 26 sank.

When the *Beagle* was launched in 1820, she was not needed by the navy, so she was held "in ordinary" (in reserve) without any masts or rigging for the next 5 years. In 1825 she was transferred to the Hydrographic Office and refitted as a hydrographic survey ship. The Hydrographic Office was sending ships all over the world to survey the oceans, islands and coasts, and draw accurate maps of them. *Beagle's* refit converted her to a bark by adding a mizzenmast toward the stern. A bark has three or more masts: the mainmast and foremast rigged with square sails, and the mizzenmast rigged with fore and aft sails. *Beagle's* armament was also reduced from 10 guns to six. Extra cabins and a forecastle were added too.

The Voyages of the *Beagle*

The *Beagle*'s first expedition as a survey ship ended in tragedy. She set sail in 1826 under the command of Captain Pringle Stokes (1793–1828) on a survey mission to Patagonia and Tierra del Fuego. Accurate charts of the South American coasts and waters were needed because Britain was developing trading relations with South America. When the *Beagle* reached Cape Horn in 1828 Captain Stokes was overwhelmed by stress and descended into a deep depression. He shot himself and died soon after. The crew brought the ship to Montevideo, where Lieutenant Robert FitzRoy (1805–65) took command.

When the *Beagle* returned to Britain, she was paid off and should have gone into reserve, but one of the ships earmarked for the next expedition to South America, HMS *Chanticleer*, was found to be in such poor condition that the *Beagle* was called back into service. However, the *Beagle* was in fairly poor condition too, so she needed a second refit before the voyage. This time, the main deck was raised by 8 inches (20 cm), the hull was sheathed in a fresh skin of planking and copper and lightning conductors were fitted. FitzRoy was given command again. For this expedition, the *Beagle* would be heading back to South America to complete the survey started by the first expedition, and then circumnavigating the globe. One aim was to plot a complete circle of longitude, so precise navigation was vital. FitzRoy wanted to replace the ship's iron guns with brass to reduce their magnetic effect on her compass, but the Admiralty refused to pay for this. FitzRoy went ahead anyway and paid for the new guns himself. The ship also carried 22 chronometers for the accurate calculation of longitude.

FitzRoy could see the wider scientific potential of the expedition, especially the possibility of collecting plant and animal specimens from remote and unknown parts of the world. He needed a naturalist for this. When he made inquiries, a young man named Charles Darwin (1809–82) was suggested. However, Darwin's father was totally opposed. He thought the expedition was a pointless diversion that would delay Darwin from settling down to more serious pursuits. But in the end he relented and allowed his son to join the expedition.

BELOW: *The* Beagle *was commanded by Captain Robert FitzRoy. FitzRoy, a pioneering meteorologist, invented the term "forecast" to describe a weather prediction.*

> NEVER, I BELIEVE, DID A VESSEL LEAVE ENGLAND BETTER PROVIDED, OR FITTED FOR THE SERVICE SHE WAS DESTINED TO PERFORM, AND FOR THE HEALTH AND COMFORT OF HER CREW, THAN THE *BEAGLE*.
>
> **Captain Robert FitzRoy (just before the *Beagle*'s second voyage, with Charles Darwin on board)**

The *Beagle* put to sea from Plymouth twice at the end of 1831, but each time she was beaten back to port by storms. She finally got under way on December 27. Darwin was instantly seasick. In the Cape Verde Islands, he was fascinated by cuttlefish that could change color at will. He also found a layer of ground containing seashells that was 45 feet (14 m) above sea level. He realized that this land must have been under the sea in the past, but he wondered how that could be. By February, the *Beagle* had reached Brazil, where Darwin went on long walks through the rain forest collecting samples of plants, insects and animals while the *Beagle* carried on with its survey work along the coast. In Patagonia, Darwin discovered huge fossilized bones. Most of them belonged to animals that were completely unknown at that time.

The *Beagle* reached Cape Horn in December and then visited the Falkland Islands. Darwin noticed that the fossils he found on the islands were different from those on the mainland. This inspired him to carry out comparative studies of all his finds in the various locations. This was the key decision that made his later work on evolution possible. The *Beagle* continued up the South American coast and Darwin took every opportunity to go ashore and collect more specimens. Whenever possible, he shipped his specimens back to England. His fourth shipment, in October 1832, included 200 animal skins, mice, fish, numerous insects, rocks, seeds and a large collection of fossils.

ABOVE: *The notes Darwin took on board the* Beagle *described the native people he encountered, including the indigenous Patagonian people, seen here in an illustration from FitzRoy's account of the voyage.*

RIGHT: *Darwin, here aged 40, was only 22 years old when he was offered a place on board HMS* Beagle.

> THE VOYAGE OF THE *BEAGLE* HAS BEEN BY FAR THE MOST IMPORTANT EVENT IN MY LIFE.
> **Charles Darwin**

ABOVE: *At the mouth of the Río Santa Cruz in Patagonia, the* Beagle *was grounded at low tide and her hull was inspected for damage. She was refloated on the next tide and continued onward.*

BELOW: *The* Beagle *carried 22 chronometers for navigation on its famous second voyage, including this model manufactured by Hugh Pennington.*

After making another trip down the South American coast and around the Falkland Islands, the *Beagle* was hauled out of the water onto the shore at the mouth of the Río Santa Cruz on April 13 to check on the state of her hull. While she was grounded, Captain FitzRoy, Darwin and other members of the crew explored the Río Santa Cruz valley. The weather was so cold that their guns froze. Darwin noticed definite layers in the ground at cliff faces, reinforcing his belief that life on Earth was continually changing, albeit very slowly, which contradicted the widespread view at that time that Earth was God's creation and therefore perfect and in no need of change over time.

In May the *Beagle*'s crew surveyed the Magellan Strait, and in June they sailed into the Pacific Ocean. Between June and August they worked their way north along the coast and Darwin went on several long treks inland. While he was in Chile, he experienced a strong earthquake that demonstrated how land can be thrust upward, leaving a shell-strewn shore high and dry.

Then there was an incident that threatened to rob Darwin of the most important part of the expedition. Captain FitzRoy had a disagreement with the Admiralty and resigned. The Admiralty's response was to order the new commander to terminate the expedition and return to England. If this had happened, Darwin would never have visited the Galápagos Islands. Fortunately, FitzRoy was persuaded to resume command and continue with the expedition.

HMS BEAGLE

The *Beagle* sailed up and down the coast of Chile and Peru completing her surveys until September 1835 and then headed for the Galápagos Islands. While the *Beagle* and her small boats wove their way around the islands, surveying the coasts, Darwin explored several of the islands and discovered their giant tortoises and iguanas. He noticed variations in the lizards, tortoises, birds and plants he found on the different islands. In particular, he noticed that the small finches on each island seemed to have a different shape of beak, as if "one species had been taken and modified for different ends."

On January 12, 1836, the *Beagle* arrived at Port Jackson, Australia, where Darwin was astonished by the strange creatures he saw. The *Beagle* made her way around the southern coast of Australia and left homeward-bound on March 13. She arrived at Cape Town in June and then headed up the west coast of Africa. FitzRoy, worried that he might have made incorrect survey readings at San Salvador, ordered a detour to the east coast of South America to recheck the work. The Beagle finally docked at Falmouth, England, on October 2, 1836, nearly 5 years after her departure.

Darwin published his thoughts and observations on geology, but he published nothing about his theory of evolution. When he discussed his ideas with friends, they were unconvinced. By then he was married and his wife, Emma, was a committed Christian. Not wishing to offend her, and worried about the wider public response, he filed a 50,000-word manuscript of a book about evolution with his private papers, together with a note asking for it to be published after his death.

Everything changed when Alfred Russel Wallace (1823–1913), who had independently arrived at his own theory of evolution, published a scientific paper

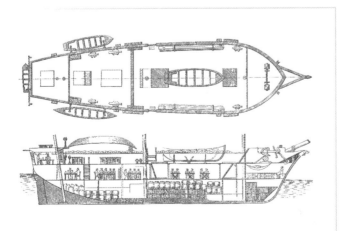

LEFT: *HMS* Beagle, *a Cherokee-class brig-sloop, was one of the last ships to be built at the Woolwich Dockyard (founded by King Henry VIII in 1512) before its closure in 1869.*

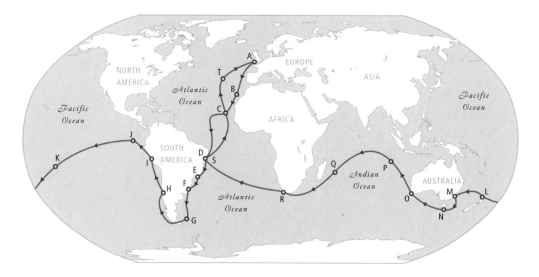

ABOVE: *The Beagle's survey expedition was expected to take 2 years. In fact, it was almost 5 years before the Beagle returned to Britain.*

A   Plymouth
B   Tenerife
C   Cape Verde
D   Bahia
E   Rio de Janeiro
F   Montevideo
G   Falkland Islands
H   Valparaíso
I   Callao / Lima
J   Galápagos
K   Tahiti
L   New Zealand
M   Sydney
N   Hobart
O   King George Sound
P   Cocos (Keeling) Islands
Q   Mauritius
R   Cape Town
S   Bahia
T   Azores

about it in 1855. Darwin was unconcerned, but some of his friends were worried that Wallace might publish a more complete theory first and get all the credit. In 1858, Wallace did indeed write just such a paper and sent it to Darwin, who was shocked by the contents because Wallace's theory was so similar to his own. However, Darwin was soon dealing with the tragic death of his son from scarlet fever. His friends stepped in and offered Wallace's paper for publication with Darwin's own theory of evolution added to it. When Darwin felt able to return to work, he wrote the book *On the Origin of Species*, which he might never have written if it had not been for Wallace.

The *Beagle*'s Last Days

The *Beagle* went on to make a third scientific expedition between 1837 and 1845. She sailed to Australia, where the first complete survey of the coast was made. On her return to Britain, her masts and rigging were removed and she was moored in the Essex marshes at Havengore Creek. She became *WV7* (*Watch Vessel 7*), a lookout post for coast guards hunting smugglers. When cottages were built nearby to house the guards, *WV7* was no longer needed and so in 1870 she was sold for scrap. There are no records of the ship after that, but some of her hull may still exist. Archaeologists using ground-penetrating radar have found a filled-in dock under a marsh in Essex. The radar images revealed the shape of a ship buried under 12 feet (3.6 m) of mud. The archaeologists think it may be the remains of the *Beagle*.

Darwin died on April 19, 1882. His funeral was held at Westminster Abbey in London and he was buried under the abbey's floor next to the astronomer Sir John Herschel (1792–1871) and near the grave of Sir Isaac Newton (1642–1727).

AMISTAD

When a group of slaves revolted and took control of the ship transporting them in the 1830s, their subsequent trial became the focus of international attention, soured relations between the United States and Spain for a generation and advanced the abolitionist movement.

TYPE: two-masted schooner

LAUNCHED: Baltimore, Maryland, ca. 1836

LENGTH: 120 ft (37 m)

TONNAGE: unknown

CONSTRUCTION: wood, carvel planking

PROPULSION: gaff-rigged sails on two masts and bowsprit

On August 26, 1839, the crew of the U.S. Revenue Cutter *Washington* stumbled upon something that looked suspicious to them. They'd spotted a black schooner anchored at the eastern tip of Long Island, New York. They could see people on the shore near the vessel. *Washington* dispatched a squad of armed officers in a boat to investigate. Most of the men on board the suspicious vessel were African, but the boarding party discovered two white men, Spaniards called Pedro Montes and Jose Ruiz, who immediately asked for the *Washington*'s protection. One of the Africans, called Cinque, tried to escape, but was captured.

The mystery vessel was the *Amistad*. She had been built in about 1836 as a schooner called *Friendship*. When she was bought by a Spanish owner, she was renamed — in Spanish *amistad* means "friendship." She was a type of schooner known as a Baltimore clipper. These were small, fast sailing ships for trade around America's coasts and in the Caribbean. They generally carried high-value, perishable cargoes that had to be transported quickly. *Amistad* worked around Cuba and other islands in the Caribbean, mainly carrying cargoes for the sugar industry. Occasionally, she also transported slaves.

Washington took *Amistad* under tow to New London, Connecticut. The U.S. government seized the ship's cargo, including 53 slaves, who were considered to be part of the cargo and therefore salvage property of the United States. It quickly became clear that the slaves had rebelled and taken control of the ship. They were imprisoned, charged with piracy and the murder of the ship's captain.

LEFT: *Joseph Cinquez, or "Cinque," led the* Amistad *revolt. When the slaves had taken control of the ship, he is quoted as saying, "I am resolved it is better to die than to be a white man's slave."*

ABOVE: *The* Freedom Schooner Amistad *is a replica of the* Amistad, *a type of fast two-masted schooner known as a Baltimore Clipper. These ships were popular cargo vessels on the eastern seaboard of the United States.*

Rebellion and Imprisonment

District judge Andrew T. Judson heard the case against them. Montes and Ruiz described what had happened. In Havana, Cuba, 53 slaves were put on board the *Amistad*, which set sail for another Cuban port, Puerto Príncipe. On the fourth night at sea, the slaves managed to break out of their irons, get up on deck and kill the ship's captain, Ramón Ferrer. Two crewmen escaped in a boat and another two, Montes and Ruiz, were spared by the slaves, because they were needed to sail the ship. The two crewmen were instructed to sail to Africa. They sailed eastward during the day, but at night they turned the ship around and sailed west. They zigzagged up the eastern seaboard for 6 weeks until they reached Long Island.

The judge referred the case for trial in the Circuit Court, where all federal criminal cases were heard at that time. The "Amistads," as the slaves were known, attracted thousands of people to the prison at New Haven where they were held. Their jailer charged members of the public a shilling to see them.

When abolitionists heard about the Amistads, they formed a committee to help prepare a legal defense. When the Spanish government learned of the case, it contended that the United States had no jurisdiction, because the *Amistad* was Spanish-owned and the slaves originated from Cuba (then Spanish territory). Spain demanded that the ship should be handed back to its owner and the slaves returned to Cuba. The U.S. president, Martin van Buren, was sympathetic to Spain's demands.

The case began in Hartford on September 14, 1839. The District Attorney asked the judge, Smith Thompson, to refer the case to the president as it had an international dimension. For the defense, Roger Baldwin argued that the United States should not act as a "slave-catcher" for other governments. Judge Thompson decided that the offenses the men were charged with had occurred in international waters and so his court had no jurisdiction. He sent the case back to the District Court to determine who owned the slaves. By then, a translator had been found to question the slaves and the story they told completely changed the case.

ABOVE: *The* Amistad *revolt and the subsequent court case were widely covered in the newspapers of the day, sometimes with fanciful and sensational illustrations.*

BELOW: *The U.S. president, Martin van Buren, was sympathetic to Spain's claim on the* Amistad *and its cargo, but the case went against him again and again as it progressed through the courts.*

Slaves or Free Men?

The men said they had recently been captured in Africa and transported from there to Cuba. By 1839, the importation of slaves to the United States was illegal, although domestic slavery continued. International treaties also banned the importation of slaves to Spanish territories. However, slavers continued to smuggle in slaves illegally, using forged documents and with the connivance of corrupt officials. Once the slaves were within U.S. or Spanish territories, they could be bought and sold legally.

The *Amistad* men said they had been seized in Mendiland (in present-day Sierra Leone) by African slavers, who took them to Lomboko, a slave port, and sold them to a Portuguese slave trader. He loaded up to 600 of them aboard a notorious purpose-built slave ship called the *Tecora*. The men were packed tightly together, naked, manacled and chained. The conditions on the ship were so appallingly awful that nearly a third of them died during the voyage to Cuba. Men who chose death by starvation were flogged into submission. Those who survived the voyage were bought by Montes and Ruiz in a slave market. Some of them were loaded aboard the *Amistad*. Cinque, the man who had tried to escape on Long Island, was the first to get loose and begin the fight for freedom.

The civil case to determine the ownership of the slaves began on November 19, 1839, in Hartford. Confident of a verdict in his favor, President van Buren stationed a ship, the *Grampus*, at New Haven, ready to spirit the Amistads away to Cuba before an appeal could be lodged. However, after hearing evidence supporting the men's claim that they had been transported illegally from Africa, Judge Judson ruled that the men were born free and had been kidnapped in violation of international law. President van Buren was "greatly dissatisfied" by the verdict. The government appealed to the Circuit Court, but lost again. They appealed once more, this time to the Supreme Court, where they were confident of success because a majority of the nine justices were, or had been, slave owners.

Former president John Quincy Adams was persuaded to speak on behalf of the Amistads. On March 9, 1841, the Supreme Court found in their favor, accepting that they were kidnapped Africans who were now entitled to their freedom. The Spanish government was furious and (unsuccessfully) sought compensation for the loss of the *Amistad* and her cargo for the next 20 years, giving up its claim only when Abraham Lincoln was elected president. In November, a ship called the *Gentleman* was chartered to return the 35 surviving Amistads to Africa.

After the case, the *Amistad*, which had been moored at New London for 18 months, was auctioned. She was bought by Captain George Hawford, who renamed her *Ion* and used her to transport fruit, vegetables and live animals. He sold her in 1844, after which there is no further record of the ship or her fate.

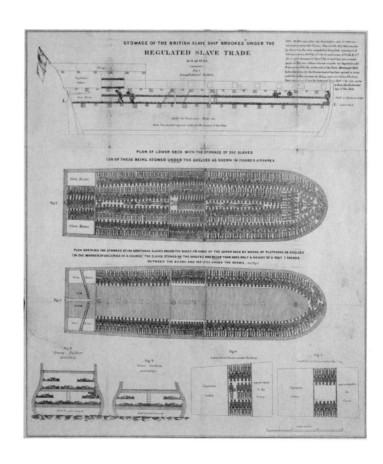

RIGHT: *The* Amistad *slaves were brought to the Caribbean on board a notorious slave ship called the* Tecora. *They were packed together so tightly, as this illustration of the British slave ship* Brookes *shows, that many of them died during the voyage.*

SS GREAT BRITAIN

The *Great Britain* represented a major leap in ship construction. When she was launched in 1843, she was the biggest iron-hulled ship that had ever been built, the largest vessel afloat by far, the first large screw-propelled ship, and the first transoceanic ship designed to operate to a schedule — the first ocean liner. She marked the beginning of the end of large wooden ships.

TYPE: passenger steamship

LAUNCHED: Bristol, England, 1843

LENGTH: 322 ft (98 m)

TONNAGE: 3,018 long tons (3,066 metric tons) displacement

CONSTRUCTION: riveted wrought-iron plates on wrought-iron frames

PROPULSION: 1,000 hp (750 kW) steam engine driving a single screw propeller plus sails on up to six masts

The SS *Great Britain* that was launched by Queen Victoria's consort, Prince Albert, on July 19, 1843, was not the ship its owners had planned to build. The ship the Great Western Steamship Company's directors had in mind was a large wooden-hulled vessel propelled by side-mounted paddlewheels, a sister ship to the SS *Great Western*. But the man in charge of the project, Isambard Kingdom Brunel, one of the most innovative engineers of the 19th century, changed the initial design dramatically by incorporating the latest developments in shipbuilding.

The ship was to be built in Bristol, a city on the River Avon in England. When an iron-hulled ship visited Bristol, Brunel sent two colleagues to investigate it. Based on their reports, he decided that the *Great Britain* should have an iron hull. And when the screw-propelled ship *Archimedes* came to Bristol, Brunel took the opportunity to check out its performance and then decided that the *Great Britain* should have a propeller instead of paddlewheels.

There were good technical reasons for both changes. An iron hull would be cheaper, lighter and stronger than wood, and also resistant

GREAT WESTERN

The *Great Britain*'s predecessor, the SS *Great Western*, was the world's biggest passenger ship when she was launched in 1838. Her designer, Isambard Kingdom Brunel, having built the Great Western Railway, suggested that the company should extend the line across the Atlantic to New York. Then he designed the great ships that would carry the passengers; *Great Western* was the first. Her design was quite conventional for the time: she was an oak-hulled, four-masted paddle-steamer. She was successful in service until retired and broken up in 1856.

ABOVE: *Brunel's SS* Great Britain *was the first large ocean-going vessel to have both an iron hull and a screw propeller.*

to rot and worm damage. Iron hulls are also stiffer than wood, so they can be built bigger. Screw propulsion used smaller and lighter machinery than paddlewheels, improving fuel economy and creating more space for carrying passengers and cargo. Screw propulsion also performed better in heavy seas and the machinery was mounted lower in the ship, improving stability. By the end of 1840, Brunel had persuaded the Great Western Steamship Company's directors to accept his design changes, which included a significant increase in the size of the ship.

The *Great Britain* was the first large ship to combine an iron hull with screw propulsion. She was also fitted with sails on six iron masts. The masts were hinged so that they could be lowered to reduce air resistance when the ship was steaming. Inside the hull, there were two passenger decks with two cargo decks below them, divided into two sections fore and aft of the massive engines and boilers, which sat amidships.

Trapped!

When the ship was launched, she was due to be towed to the Thames for fitting out. However, she was too big to negotiate the lock gates between the dock and the Bristol Channel; she was trapped. The harbor authorities had to be persuaded to undertake costly modifications before she could leave. Even after the modifications, the huge ship got jammed in one of the gates, necessitating the removal of some of the lock's stonework to let her finally pass through to the Avon and the sea beyond.

On July 26, 1845, 2 years after the great ship was launched, she departed on her maiden voyage from Liverpool to New York. She could carry 360 passengers, but there were only a few dozen on board. She suffered from a series of technical and performance problems. Her speed was disappointing, the ship taking about 14 days to cross the Atlantic. Brunel modified the propeller for more speed, but on the next crossing she lost three of her six propeller blades. After repairs, she lost four blades on the return journey. The passengers and crew were also dismayed to find that the ship rolled badly, even in calm weather. The propeller and rigging were replaced, and bilge keels were added to resolve the stability problems. One of the six masts was removed too. The owners not only had to pay for the modifications, they also lost income from the ship while she was out of service.

Her next season looked more promising until she ran aground in Dundrum Bay on the northeast coast of Ireland. It was nearly a year before she was refloated. By then, her owners had decided that they could not spend any more money on her and sold her for a fraction of her value.

The new owners gave her a refit and strengthened her hull. Her engines were replaced by more modern, smaller engines with smaller, high-pressure boilers, and her five masts were further reduced to four. After just one transatlantic trip she was sold again. This time, her owners planned to use her on the England to Australia route. They increased her passenger capacity from 360 to 730 and reduced her four masts to three. She caused a sensation on arrival in Australia;

RIGHT: *The* Great Britain's *bow is adorned with the royal coat of arms of the United Kingdom, with a lion on one side and a unicorn on the other side.*

GREAT EASTERN

After the *Great Western* and *Great Britain*, Brunel designed the SS *Great Eastern*. She was a monster of a ship, six times the size of any other ship afloat. She had sails, paddlewheels and screw propulsion. Her launch on November 4, 1857, went disastrously wrong. She was so long that she had to be launched sideways, but she stuck fast on the slipway and refused to slide into the water. Even steam winches and hydraulic rams couldn't shift her. It took 3 months to get her into the water. She was designed to carry 4,000 passengers, but only ever carried a fraction of that number. She was sold and converted to a cable-laying ship for the growing international telegraph industry. A reconversion back to a passenger liner failed and she was eventually scrapped in 1889–90.

ABOVE: *Brunel's* Great Eastern *was such a massive ship that her size and weight were unmatched by any other vessel for decades. Brunel is said to have referred to the ship as the "Great Babe."*

thousands of people paid a shilling each to see her. The *Great Britain* stayed on the Australia run for 30 years, apart from two spells as a troop transport ship during the Crimean War and the Indian Rebellion of 1857. Her longest-serving captain during the Australian years was John Gray, who disappeared in mysterious circumstances. He vanished from the ship during a return voyage from Australia in 1872. He had been a very popular captain with his crews and passengers.

Coming Home

In 1882, the *Great Britain* was converted to carry coal. Four years later, she put into Port Stanley in the Falkland Islands with storm damage, but it was found that she would be too expensive to repair. She stayed there and was used as a coal bunker until the 1930s, when she was scuttled in shallow water and abandoned. Finally, in 1970, the great ship was loaded onto a pontoon and towed back to her birthplace, the Great Western Dockyard in Bristol, where she was restored and put on public display.

HMS RATTLER

Navies were slow to adopt steam power, because the paddlewheels that propelled early steamships were too easily damaged in battle and they also cut down the amount of hull space available for mounting guns. Then inventors and engineers developed a new way to move a ship through water — the screw propeller.

TYPE: screw sloop

LAUNCHED: Sheerness Dockyard, England, 1843

LENGTH: 185 ft (56.4 m)

TONNAGE: 880 long tons (894 metric tons) displacement

CONSTRUCTION: wooden hull, carvel planking

PROPULSION: barquentine sail plan and 200 hp (150 kW) steam engine driving a single screw propeller

BELOW: *David Bushnell had equipped his submersible, the* Turtle, *with propellers long before the* Rattler *trialed the use of screw propellers.*

The Royal Navy recognized the advantages of steam power in terms of performance but because of its unsuitability for large warships, its use was limited to auxiliary vessels and small gunboats that did not engage in combat with other ships. The navy finally became interested in adopting steam power for warships when it became aware of a small experimental steamboat called the *Francis Smith* designed by Francis Pettit Smith. Instead of paddlewheels, the boat's steam engine turned a single screw propeller.

Using a screw-shaped device for propulsion was not a new idea in the 19th century. It dates back more than 500 years to Leonardo da Vinci's design for an airscrew — a helicopter-like machine. Various inventors, scientists and engineers proposed the use of screw-shaped propellers in water. Bushnell's *Turtle*, a primitive diving craft built in 1776, used hand-cranked screws for propulsion and to fine-tune its depth. Robert Fulton experimented with propellers in the 1790s. In the early 1800s there was a flurry of patent applications dealing with propeller design; one of them was filed by Francis Pettit Smith. His boat's ability to keep going during a storm that would have challenged a paddle-steamer rekindled naval interest in steam power.

The British Admiralty asked Smith to build a full-size ship to test the technology. The result was the SS *Archimedes*. After some initial

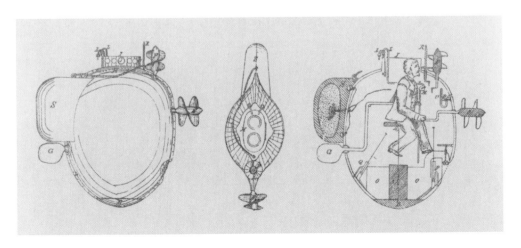

ABOVE: *The* Rattler *and the* Alecto *competed against each other in the world's most famous tug of war in 1845, comparing the performance of screw propellers and paddlewheels.*

technical problems, including a boiler explosion and a broken crank-shaft, the admiralty trialed *Archimedes* against the fastest paddlewheel vessels and found her performance to be at least equal to theirs. She was then taken on a voyage around Britain so that the naval officers on board could test her in a variety of weather conditions. After these tests, the Great Western Steamship Company borrowed her for trials, inspiring Isambard Kingdom Brunel to adopt screw propulsion for his SS *Great Britain*. Meanwhile, the navy built its first screw-propelled warship. She was originally intended to be a conventional warship under sail, called *Ardent*, but in mid-construction the admiralty ordered her to be converted to screw propulsion — and they changed her name to *Rattler*.

ABOVE: *The first propeller fitted to the SS* Archimedes *looked more like a short section of an Archimedes' screw than a modern multibladed propeller.*

Propellers Versus Paddlewheels

Rattler was a nine-gun sloop fitted with a 200 hp (150 kW) steam engine driving a single screw propeller. She was operated by a crew of 180 officers and men. After her launch on April 13, 1843, at Sheerness Dockyard, she spent the next 2 years in trials. The navy experimented with different shapes and sizes of propeller to find the most efficient design. The most famous of her sea-trials took place in March and April 1845, when *Rattler* competed with the paddle-steamer HMS *Alecto*.

The two ships were of comparable size, weight and engine power. First, *Rattler* raced against *Alecto*. Over an 80-mile (130-km) course off the east coast of England, *Rattler* finished in 8 hours 34 minutes, more than 23 minutes ahead of *Alecto*. Then, over a 60-mile (100-km) course with strong winds and heavy seas to contend with, *Rattler* finished 40 minutes ahead. For the last of the 12 trials the two ships were tied stern to stern, facing in opposite directions. *Alecto* reached full power first and towed *Rattler* along for several minutes. Then, when *Rattler* reached full steam power, she slowed *Alecto* to a halt and then towed her backward at up to 2.5 knots (2.9 mph or 4.6 km/h). *Rattler* was such a clear winner of the trials that all Royal Navy ships were fitted with propellers from then on. Further trials in 1846 between the paddle-steamer *Basilisk* and the screw-propelled *Niger* produced similar results.

When the trials were over, *Rattler* entered service with the navy's Experimental Squadron, whose job was to test new hull shapes, propulsion, armaments, etc. in service. One of her first tasks was to tow the ships *Erebus* and *Terror* to Orkney on the first leg of the Franklin expedition that subsequently vanished in the Canadian Arctic. On October 3, 1849, while on antislavery patrol off West Africa, she captured the Brazilian slave brigantine *Alepide*. In 1852–53, she served in the Second Anglo-Burmese War. In 1855, she helped to suppress the activities of Chinese pirates preying on international merchant shipping. She was finally scrapped on November 26, 1856.

JOHN ERICSSON

John Ericsson, one of the inventors of screw propulsion, had an extraordinary life. He was born in Sweden in 1803. After a spell of service with the Swedish army, he moved to England in 1826 with high hopes of selling a heat engine he had designed. But the wood-burning engine didn't run well on the main fuel used in Britain — coal. Undeterred, he invented improvements for steam engines. He built several locomotives, although they weren't commercially successful. He lost so much money on these ventures that he spent some time in a debtors' prison. It was then that he began the work on propellers that would take him to the United States. After building the USS *Princeton*, he built the first iron steamboat (the *Iron Witch*), invented a hot-air engine and designed an ironclad warship, the USS *Monitor* (see pages 102–105). Ericsson died in New York on March 8, 1889, at the age of 85 after a very eventful life.

Meanwhile, in the United States . . .

Francis Pettit Smith was not the only person engaged in the development of screw propulsion, and *Rattler* wasn't the only significant ship in this respect. Six weeks after Francis Pettit Smith patented his propeller in 1836, another inventor, John Ericsson, filed his own patent for a propeller. After some small-scale experiments he built a 40-foot (12-m) boat driven by two screw propellers and tested it successfully on the River Thames. At that stage, however, the admiralty was yet to be convinced of the suitability of the screw propeller for warships, so in 1839 Ericsson moved to the United States where he built the U.S.

RIGHT: The explosion of a gun on board Ericsson's propeller-powered warship, the USS Princeton, killed several guests and marred the ship's success.

Navy's first propeller-driven warship. This vessel, the USS *Princeton*, was launched in September 1843. Although *Princeton* was launched a few months after *Rattler*, she was the first to be commissioned and so has a claim to be the first screw-propeller warship.

Unfortunately, she is remembered more for a terrible accident. With the president and several members of the government on board, the ship demonstrated her weaponry by firing her guns. One of the guns exploded, killing six people including the secretary of the navy and the president's valet. Another 20 people were injured. The accident prompted a review of gun manufacturing methods.

The *Princeton* served with the U.S. Navy until 1849, when her timbers were found to be in a very poor condition and she was retired to be broken up. As a result of Smith's and Ericsson's pioneering work, the screw propeller quickly became the standard naval propulsion method. In 1850, the French launched their first steam-powered, screw-propelled warship, *Napoléon*, and other navies quickly followed.

RIGHT: From its origins as a crude twist of iron plate, the propeller quickly developed in design and size into the giant multi-bladed propellers that powered the great ocean liners of the 20th century.

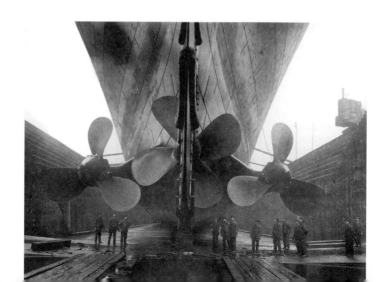

H M S   R A T T L E R

AMERICA

As the use of sailing ships for trade and warfare went into decline in the 19th century, the idea of sailing for pleasure grew in popularity. The first major international yacht race was held off the south coast of England in 1851. The winner, a yacht called *America*, gave the race and its trophy the name they are still known by today — the America's Cup. That first winner in 1851 influenced yacht design for years to come, and current America's Cup yachts continue to push designers to the limits of yacht technology.

TYPE: gaff schooner

LAUNCHED: New York, 1851

LENGTH: 101 ft (31 m)

TONNAGE: 100 tons

CONSTRUCTION: wood (white oak, locust, cedar and chestnut)

PROPULSION: sail

The Royal Yacht Squadron in Britain offered its "One Hundred Sovereign Cup" for the winner of a race around the Isle of Wight, a large island off the south coast of England, in 1851. At the same time in the United States, Commodore John Cox Stevens and a group of friends from the New York Yacht Club built a yacht with the intention of taking it to Britain to win prize money in competitions.

The designers they chose for the job were James Rich Steers and his brother, George. Their design resembled the hull shape of a clipper ship, with a sharper, more concave bow and with the widest part of the boat farther back than in a traditional racing yacht. She was rigged as a schooner with steeply raked masts. Pilot boats like this raced against each other to meet oceanic vessels and put pilots on board to guide the big ships into and out of ports; the fastest boats secured the most business. Their captains and crews were skilled in fast sailing and maneuvering through the shallows and channels close to shore. One of them, Captain Richard Brown, was selected to command *America*.

RIGHT: *Fitz Hugh Lane's oil painting shows the racing yacht* America *winning the "International Race" in 1851, taking the trophy that was named after her, the America's Cup.*

Ready to Race

In July 1851, *America* arrived in Le Havre in Normandy, France, where she was repainted before leaving for the Isle of Wight. The noted American newspaper editor and congressman, Horace Greeley, strongly advised the yacht's owners not to race in England, because they would surely be beaten and this would reflect badly on the United States. But they were already committed to competing. The British treated the new American yacht as a curiosity and so different from the British yachts that she could not be a serious contender. The *Illustrated London News* described her as "artistic, but rather a violation of the old established ideas of naval architecture." Her owners had difficulty finding anyone to race against. The challenges they offered were not taken up. Finally, they were invited to race on the last day of the regatta. The race held on that day was traditionally a members-only event, but to accommodate *America* it was opened to entries from yacht clubs of all nations.

The rule that each yacht must have a single owner was also suspended for *America*, and she was allowed to "boom out" her sails — hold the sails out with poles to help them fill with air when running downwind.

On August 22, 1851, with Queen Victoria watching from the royal yacht, 15 schooners and cutters lined up for the 53-nautical-mile (98-km) race. *America* had a poor start because of a fouled anchor. When she eventually got under way she was well behind all the other boats, but she quickly

> SHE IS ARTISTIC, BUT RATHER A VIOLATION OF THE OLD ESTABLISHED IDEAS OF NAVAL ARCHITECTURE.
>
> **Illustrated London News**

AMERICA

91

gained on them. As the boats reached the east end of the Isle of Wight, they traditionally rounded the island on the seaward side of a lightship that marked the Nab Rocks, but there was nothing in the printed rules to stop yachts from taking a shorter, riskier route between the island and the rocks. That is precisely what *America* did, enabling her to pass the other yachts and take the lead. She stayed in front, even when her jib boom broke and had to be replaced.

Eight and a half hours after the start, *America* finished 8 minutes ahead of her nearest rival, *Aurora*. A protest about the route *America* had taken at the Nab Rocks was lodged but

disallowed. *America*, the "curiosity," had beaten the best British yachts in their home waters at her first attempt. The next day, the queen and Prince Albert paid a visit to the boat. The commentators of the day saw more in *America*'s success than a mere yacht race victory. The *Merchant*, a London newspaper, wrote, "The empire of the seas must before long be ceded to America."

The men who had built *America* brought the trophy home and donated it to the New York Yacht Club on condition that they offer it as a challenge trophy to promote friendly competition between nations. In honor of the yacht that had won it, it was called the America's Cup. United States teams held it for the next 132 years before the first non-U.S. team won — an Australian team with the yacht *Australia II* in 1983.

Decline and Fall

America's owners sold her 10 days after the 1851 race. She was bought and sold several times by British owners over the next few years. Her success influenced British yachts and yacht designers, as have subsequent America's Cup yachts. *America* was rebuilt and, then called *Camilla*, was sold to the Confederate States of America, who used her as a blockade runner during the Civil War. When Union troops took Jacksonville, she was scuttled in Dunns Creek, north of Crescent City. The Union forces raised her, renamed her *America* and used her during

SPACE-AGE YACHTS

The latest America's Cup yachts are worlds away from *America* and her rivals in 1851. The rules governing the design of the yachts have changed repeatedly over the years to reflect advances in materials and technology, and to ensure that the yachts are closely matched for exciting racing. The latest and most advanced yachts "fly" above the water, standing on underwater wings called hydrofoils. The space-age materials the yachts are made from and the streamlined shapes of their hulls, sails and foils are so important that the yacht development teams often work closely with aircraft designers and Formula 1 racing-car designers.

BELOW: *Two America's Cup yachts skim the wave-tops during the 2013 race. These thoroughbred racing machines exceed 50 mph (80 km/h).*

the war. The yacht that was only ever intended to be a racing boat was armed with three cannons: a 12-pounder and two 24-pounders. After the war, she was used as a training ship at the U.S. Naval Academy. She took part in the America's Cup again in 1870 and finished fourth. The navy sold her in 1873 and she was rebuilt 2 years later. She was used for racing and recreational sailing until 1901, when she was largely abandoned and fell into disrepair. In 1921, she was donated to the U.S. Naval Academy, but did not receive the extensive restoration work she needed. By the 1940s, she was in urgent need of rescue. In 1942, the shed she was kept in collapsed. Three years later, she was finally scrapped and burned.

HMS CHALLENGER

Forty years after Charles Darwin's voyage in the *Beagle*, a converted British warship, HMS *Challenger*, carried out the first worldwide scientific study of the oceans. In doing so, she laid the foundations of the new science of oceanography.

TYPE: *Pearl*-class corvette

LAUNCHED: Woolwich Dockyard, England, 1858

LENGTH: 225 ft (68.7 m)

TONNAGE: 2,137 long tons (2,171 metric tons) displacement

CONSTRUCTION: wood, carvel planking

PROPULSION: full-rig square sails plus a 1,200 hp (895 kW) steam engine driving a single screw propeller

*U*ntil the 1860s, scientists believed that the deep ocean floor must be lifeless, because no light from the surface ever reached it. However, dredging off the British and American coasts disproved this when prolific marine life was found at great depths. Scientists realized that there was a new world of living creatures on the ocean floor waiting to be discovered. The *Challenger* expedition was one response to this. The expedition was the idea of Edinburgh University's Professor of Natural History, Charles Wyville Thomson. At his suggestion, the Royal Society in London asked the British government to provide a ship for an expedition to study the oceans. The government approved the expedition and made HMS *Challenger* available for it.

Challenger was a steam-assisted sailing ship, a *Pearl*-class corvette. Corvettes were, and still are, small warships. Smaller than frigates, they are typically employed in coastal patrol, fleet support and fast attack operations. The French navy was the first to describe small warships as corvettes in the 1670s and the word is thought to be a form of an older Dutch word, *corf*, meaning a small ship.

RIGHT: Challenger *spent February 1874 sailing south into the Antarctic. The crew had to contend with mountainous seas, icebergs and pack ice.*

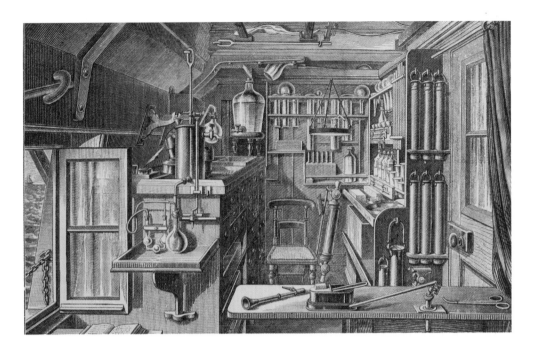

ABOVE: Challenger's chemical laboratory was in a tiny cabin measuring only 10 feet by 5 feet (3 m by 1.5 m). Every piece of equipment had to be secured in its place.

BELOW: The expedition was the brainchild of Scottish naturalist Charles Wyville Thomson, who served as its chief scientist.

Challenger was one of 10 *Pearl*-class corvettes built in the 1850s. She served in the Americas in the early 1860s before being ordered to Australia. In 1870 she was selected for the scientific survey that became known as the *Challenger* expedition. To prepare her for the task, some of her guns were removed to make more room for equipment. Extra cabins were fitted, together with a dredging platform where scientists could bring aboard samples of seabed mud and creatures from the depths. She was equipped with two laboratories fitted out with all the instruments the scientists would need.

When *Challenger* left Britain in December 1872, she had 243 officers, crewmen and scientists on board. She was commanded by Captain George Nares. Charles Wyville Thomson supervised the scientists. They headed south across the equator into the South Atlantic and then rounded the Cape of Good Hope into the southern Indian Ocean. On February 16, 1874, *Challenger* became the first steamship to cross the Antarctic Circle, although she was actually under sail most of the time; steam power was generally used only when the weather was too calm for sail. Dredging in the cold southern ocean waters brought up about 400 different species of animals, more than three-quarters of which had never been seen before. *Challenger* did not travel much farther south because her hull had not been strengthened for ice. The ship next headed for Australia and New Zealand, then sailed north through the western Pacific and crossed the ocean to the Hawaiian Islands. From there, she sailed to Cape Horn and finally

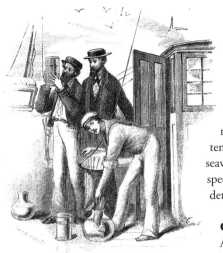

ABOVE: *Three of* Challenger's *crewmen inspect samples of marine creatures brought up from the deep, in this case medusa jellyfish.*

headed north through the Atlantic back to Britain. She arrived at Portsmouth on May 24, 1876.

As she crisscrossed the oceans, the ship stopped every 200 miles (320 km) or so to collect samples and record observations. At each of 362 sites, the water depth was measured and a sample of the seabed was taken. The water temperature was measured at different depths. Samples of seawater were taken at different depths too. The water current's speed and direction at the surface were noted, along with details of the weather.

Challenger Deep

A total of 492 depth soundings were made. One of them, at sampling site 225 between the islands of Palau and Guam in the western Pacific, turned out to be particularly deep. *Challenger* recorded a water depth of 26,850 feet (8,184 m). Nearby, in fact, was the deepest-known part of any ocean, later measured at 36,070 feet (10,994 m) deep. This spot was named Challenger Deep after the expedition. *Challenger*'s depth soundings revealed the first general map of the shape of the ocean floor, including a rise in the middle of the Atlantic. This was the first indication of what would later be identified as the Mid-Atlantic ridge, the longest mountain range on Earth, lying along the fault where two of Earth's tectonic plates are moving apart.

By the end of the 3½-year expedition, Challenger had sailed 68,890 nautical miles (79,280 miles or 127,580 km). When all the samples and observations were unloaded, more than 100 scientists pored over them. The research proved conclusively that marine life does exist at great depths all over the oceans. The report the scientists produced, presenting the results of the research, took 23 years to write and filled 50 thick volumes running to a total of 29,500 pages. The expedition's discoveries included 4,700 new species of plants and animals.

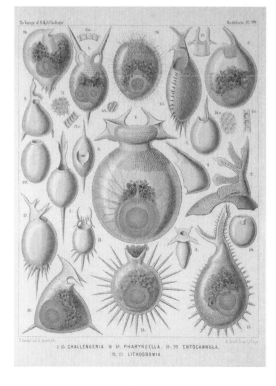

LEFT: Challenger's *scientists made copious notes and drawings detailing the samples they collected during the expedition. Their published findings occupied 50 volumes.*

THE *METEOR* EXPEDITION

Dozens of scientific expeditions and surveys by many nations followed the *Challenger* expedition. Each of them revealed a little more about the oceans, their currents, the ocean floor, seamounts and marine life. The development of sonar for the detection of submarines during World War I led to acoustic depth sounding — measuring the ocean's depth by bouncing sound waves off the seabed. Until then, ships had to stop at each sounding site and lower a weight to the seabed to measure the water depth; acoustic sounding was faster and easier. The German *Meteor* expedition (1925–27), carrying early sonar equipment, made 67,000 depth soundings compared to fewer than 500 made by the *Challenger* expedition. They proved that the rise in the middle of the Atlantic Ocean detected by *Challenger*'s scientists was actually one continuous mountain range.

Challenger's Last Days

After her historic voyage, *Challenger* was used as a Coast Guard and Royal Naval Reserve training ship at Harwich on the east coast of England. Then between 1878 and 1883 she was held in reserve before being converted to a "receiving hulk." Receiving hulks were ships that housed newly recruited sailors while they were waiting to be assigned to a ship. At that time some sailors were still being pressed into service unwillingly, so the navy had to confine them to prevent them from escaping until they were safely at sea. Receiving hulks were the answer. Some receiving hulks were also used as floating hospitals, especially if there was no shore hospital nearby. *Challenger* served as a receiving hulk on the River Medway until she was sold to a breaker in 1921. Her figurehead is the only part of her that has survived to the present day.

The many discoveries made by the *Challenger* expedition prompted other countries to begin studying the oceans too. More recently, one of NASA's Space Shuttle orbiters was named *Challenger* to commemorate the scientific achievements of the ship and her crew.

LEFT: *The German ship* Meteor *departs her home port in 1925 at the beginning of her 2-year marine survey expedition in the Atlantic Ocean.*

GLOIRE

A development in naval gunnery in the middle of the 19th century led to the most radical change in the design of warships for a thousand years. The first large warship of a major European naval power to incorporate the new features was the French battleship, *Gloire*. Her appearance triggered an arms race in battleship design.

TYPE: *Gloire*-class ironclad battleship

LAUNCHED: Toulon, France, 1859

LENGTH: 256 ft 6 in (77.9 m)

TONNAGE: 5,630 long tons (5,720 metric tons) displacement

CONSTRUCTION: wood clad in wrought-iron plates

PROPULSION: sails and a 2,500 hp (1,900 kW) steam engine

On November 30, 1853, Russian warships annihilated a Turkish fleet at the Battle of Sinope during the Crimean War. The Russian ships were armed with guns firing explosive shells. The shells blasted through the wooden hulls of the enemy ships and started fires that burned the ships down to the waterline. The major European powers quickly realized the implications for them: Their vast navies of wooden-hulled ships were immediately rendered obsolete.

The first country to react to the threat was France. During the war, both Britain and France had used floating gun platforms armored with iron plates. Now, France applied the same technology to its warships. It ordered the construction of four ironclad battleships. The weight of their armor and the bigger steam engines needed to propel them meant that the new ships would have to be bigger than their predecessors. The first, *Gloire* (*Glory*), was launched on November 24, 1859, and completed the following year. She was the world's first oceangoing ironclad battleship. Her hull was composed of 26 inches (66 cm) of wood clad in iron plate armor up to 4.6 inches (120 mm) thick. The armor extended about 6 feet (1.8 m) below the waterline. Armor like this, extending from a ship's deck to just below the waterline but not covering the whole hull, is called "belt" armor. It is designed to protect the most vulnerable part of the hull from artillery shells and torpedoes.

LEFT: *At the Battle of Sinope, Ottoman ships were blasted to pieces and set on fire by explosive shells fired by Russian warships. The vulnerability of wooden-hulled ships was immediately obvious, prompting the development of iron-hulled warships.*

ABOVE: *The French ironclad* Gloire *was the first of a new generation of warships. Rival navies were soon building their own ironclads.*

PROTECTING THE HULL

Navies discovered that their ironclad hulls needed more maintenance than wooden hulls. The iron quickly corroded in seawater and became fouled with marine life that slowed the ships down. Copper had been used to protect wooden hulls from shipworm and other marine life for the previous 100 years, but it could not be used with iron hulls. When the British frigate HMS *Alarm* was sheathed in copper in 1761, iron nails in contact with the copper dissolved; in places where the copper and iron did not touch, the iron was undamaged. The effect is called galvanic corrosion, and it can still be a problem. The U.S. Navy littoral combat ship *Independence*, launched in 2008, suffered from serious corrosion problems, reportedly because of a galvanic reaction between steel engine parts and her aluminum hull.

Sail and Steam

In common with other 19th-century warships, *Gloire* used both sail and steam power. Her hull housed a 2,500 hp (1,900 kW) steam engine, amply supplied with steam by eight boilers. The engine drove a single screw propeller. Her sail plan was changed several times during her short career. Her original three-mast barquentine rig comprised a square-rigged foremast, with fore-and-aft-rigged main and mizzen masts. Later, this was converted to square-rigged throughout, and finally to all fore-and-aft sails.

The next two ships of the *Gloire* class, *Invincible* and *Normandie*, were built in the same way as *Gloire*. The fourth, *Couronne*, was different: She was France's first iron-hulled ironclad. *Couronne*'s hull was composed of armor plating over 4 inches (100 mm) of teak, backed by an iron lattice over another 12 inches (300 mm) of teak fixed to an iron hull. This complex composite hull construction proved to be more effective than the simpler hulls of the earlier ships and was adopted for the construction of all French warships.

A WARRIOR MADE OF IRON

The Royal Navy had been the world's most powerful navy since the end of the 17th century — and it intended to remain so. Its response to the French ironclad *Gloire* was HMS *Warrior*. Launched in 1860, *Warrior* was significantly bigger and heavier than *Gloire*. She was also faster and more heavily armed. And she was built differently — while *Gloire* was essentially a wooden-hulled ship covered with iron, *Warrior* was an iron-hulled ship backed by wood. She was the first armor-plated, iron-hulled warship. Iron armor plating 4.5 inches (114 mm) thick was backed by 18 inches (45 cm) of teak, all of which was bolted to the ship's 1-inch (25-mm) thick iron hull. *Warrior* and her sister ship, *Black Prince*, instantly became the world's most powerful warships. They were so formidable that the Royal Navy did not build any more wooden-hulled warships. The U.S. Navy quickly followed suit, launching its first iron-hulled warship, the USS *Michigan*, in 1863. Iron hulls were quickly replaced by lighter and stronger steel hulls in the 1870s. The French were the first again, with *Redoutable* in 1876.

ABOVE: *HMS* Warrior *was the world's fastest, biggest and most powerful warship upon her 1860 launch. She showed that Britain intended to maintain its supremacy at sea.*

LEFT: *Although 19th-century wooden-hulled ships were typically armed with 32-pounder guns, Warrior's gundeck bristled with 68-pounders and 110-pounders.*

A Cat and Mouse Game

The introduction of ironclads required the development of bigger and more powerful naval guns to penetrate their armor plating. The guns soon became too heavy for seamen to haul back inside the ship for loading. In another development, smoothbore guns were replaced by guns with rifled barrels; spiral grooves inside the barrel made the shell spin. Rifling made the guns more accurate and their shells more penetrating, but it also made muzzle-loading impractical. As a result, breech-loading guns began to replace muzzle-loaders. The cat and mouse game of more powerful guns triggering the development of bigger battleships with thicker armor, leading to further improvements in gunnery, continued well into the 20th century.

Gloire's armament included several Paixhans guns. Invented by the French general and artillery officer Henri-Joseph Paixhans in 1823, they were the first naval guns designed to fire explosive shells. They are also known as canons-obusiers (shell-gun cannons). Explosive shells used in land warfare were usually fired high in the air to fall upon the enemy. Naval gunnery required projectiles to be fired faster by high-powered guns along a flatter trajectory into the side of an enemy ship. Cannons firing cannonballs had done this job very effectively for hundreds of years. It was too dangerous to fire explosive shells in naval guns because of the risk of early detonation, until Paixhans developed a fuse that delayed a shell's detonation until it hit the target. It was Paixhans guns that were used by Russia at the Battle of Sinope, prompting the development of *Gloire* and all the other ironclad warships that followed her.

Despite *Gloire*'s advanced technology, she was relatively unsuccessful. She rolled badly at sea, a severe shortcoming for a ship that sat so low in the water — her gun deck was less than 6 feet (2 m) above the waterline. She was struck from the navy list in 1879, only 19 years after her completion, and scrapped later the same year.

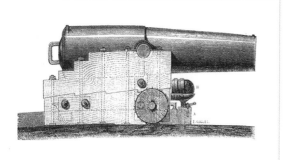

ABOVE: *A new generation of more powerful naval guns, invented by French artillery officer Henri-Joseph Paixhans, prompted navies to build bigger warships with thicker armor.*

BELOW: *HMS* Warrior's *armor had a 2 feet (60 cm) thick composite construction of iron plates over teak fixed to the ship's iron hull plates, which were themselves backed by more teak.*

USS MONITOR

During the Civil War, two ironclad warships met in battle for the first time. The USS _Monitor_ and CSS _Virginia_ pounded each other at the Battle of Hampton Roads in 1862. The event received international attention. It hastened the transition from wooden-hulled ships to ironclads and iron-hulled ships. The _Monitor_ also influenced the way later warships would be armed.

TYPE: _Monitor_-class warship

LAUNCHED: Brooklyn, New York City, 1862

LENGTH: 179 ft (54.6 m)

TONNAGE: 987 long tons (1,003 metric tons) displacement

CONSTRUCTION: ironclad hull

PROPULSION: vibrating-lever steam engine driving a single propeller

BELOW: _Engineering drawings show_ Monitor's _flat-topped shape and gun turret, and how the deck overhang protected the propeller and lower part of the hull._

$\mathcal{W}$hen the Union Navy learned that the Confederates were building an ironclad warship, they quickly ordered an ironclad of their own. In fact, they received 17 design proposals and built three of them: the USS _Galena_, _New Ironsides_ and _Monitor_. _Galena_ and _New Ironsides_ were conventional ships called "broadside ironclads." They were iron-armored copies of wooden-hulled fighting ships, with cannons firing through gunports in the sides of the ship. _Monitor_ was completely different.

Monitor's design was the work of two people. Theodore Ruggles Timby had invented a rotating gun turret in the 1840s, but it attracted little interest from the government or military until the Civil War. Timby's proposal to Lincoln's Union leadership during the war coincided with the arrival of a revolutionary design for a new ironclad warship by John Ericsson, one of the inventors responsible for the screw propeller. The two designs were combined to produce the USS _Monitor_.

Monitor was a very odd-looking shape, but it was a very functional shape. The top part was an armored wooden raft that overhung the lower part of the hull to protect it and the propeller from enemy fire and ramming. About an inch (25 mm) of wrought-iron armor protected the deck. The sides were more thickly armored with up to 5 inches (13 cm) of iron plates over 30 inches (76 cm) of wood. The freeboard (the distance between the waterline and the deck) was only 14 inches (35 cm), to present the smallest possible target to the enemy. The disadvantage of such a small freeboard was that water poured

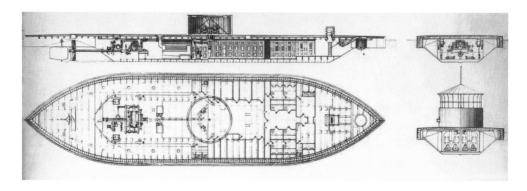

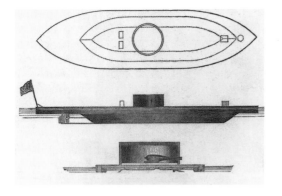

through the deck hatches if they were opened in anything but the flattest sea conditions. The only part of the ship that stood above the deck was Timby's gun turret. The turret gave *Monitor* a major advantage in confined spaces where maneuvering was difficult. With a conventional warship the whole vessel had to be turned to bring some of its guns to bear on a target, but *Monitor*'s turret could turn to point its guns in almost any direction.

ABOVE: *Designed to sit very low in the water,* Monitor *presented almost nothing for an enemy to fire at.*

BELOW: Monitor *suffered no significant damage in battle, even when she suffered direct hits, but was unable to inflict any damage on the equally impervious* Virginia.

From *Merrimack* to *Virginia*

The ship that would be *Monitor*'s opponent at the Battle of Hampton Roads, the CSS *Virginia*, was originally built as a wooden-hulled, sail-assisted steam frigate called the USS *Merrimack* (or *Merrimac*). When the state of Virginia seceded in April 1861, the U.S. Navy attempted to sail the *Merrimack* out of the Norfolk Navy Yard, but their way was blocked. Unable to get the ship out, they set fire to her before making their escape. She burned down to the waterline and sank.

The Confederate states were so desperate for ships that they raised the *Merrimack* and rebuilt her. Since nothing remained above the waterline, they had a clean sheet to rebuild her in any way — and they

ABOVE: Monitor's officers inspect dents in the ship's gun turret armor caused by rounds from the ironclad Virginia bouncing off it. The muzzle of one of her two Dahlgren guns is visible.

BELOW: This engraving from Harper's Weekly shows the attempted rescue of Monitor's crew shortly before she sank in a storm off Cape Hatteras. Behind her is USS Rhode Island.

decided to convert her to an ironclad. The design they chose was not a monitor or a broadside ironclad, but a third type called a "casemate" ironclad. The main deck where the guns were located was encased in iron armor, with sloping sides to deflect direct hits. Like the *Monitor*, she had little freeboard.

The Battle of the Ironclads

Monitor reached Hampton Roads on March 9, 1862, at the end of the first day of the battle. She slipped unseen alongside a grounded warship, the USS *Minnesota*, under cover of darkness. The next morning, as the CSS *Virginia* ironclad made for the *Minnesota* and opened fire to finish her off, the *Monitor* emerged from behind the ship and returned fire. The two ironclads pounded each other for about 4 hours. *Monitor*'s steam-powered turret malfunctioned during the battle. The crew had to keep it turning continuously, firing the guns as they came to bear on the *Virginia*. Each ship scored direct hits on the other and they were so close that they even collided on occasions as they maneuvered around each other, but neither ship suffered any significant damage. Unable to gain an advantage, both vessels withdrew.

Monitor went on to take part in the Battle of Drewry's Bluff, an attempt to reach the Confederate capital at Richmond, Virginia, on the James River, and bombard it. However, the attack failed because the river was blocked. *Monitor* was moved to the Washington Navy Yard in October 1862, for repairs to her engines. Thousands of people, including President Lincoln, came to see her. Her repairs and improvements were completed by November. Her crew received orders to make for North Carolina and join the blockade of Charleston. A storm blew up while she was under tow off Cape Hatteras, North

A CHEESE ON A RAFT
A Confederate sailor's description of the USS *Monitor* on seeing it for the first time

Carolina. With so little freeboard, the rough sea washed over her deck and flooded in through vents. Her captain ordered the towline to be cut. Two men who volunteered for this were swept off the deck and drowned. The line was finally cut and all engine power was diverted to the pumps, which failed. The water continued to rise until *Monitor* sank with the loss of 16 men who had remained on board. Another 47 who had abandoned ship were rescued by lifeboats from Rhode Island.

MONITOR'S LEGACY

Monitor was so successful that more ironclads like her were built. Many of them were used in river battles on the Mississippi and on the James River. John Ericsson, who designed the *Monitor*, went on to create a new class of monitor-type ships called the *Passaic* Class. They were bigger than the *Monitor* and had an improved hull shape with less overhang. The British Royal Navy also developed an ironclad. Called a "breastwork" monitor, it had a low armored superstructure (breastwork) built up from the deck. This eliminated the danger of flooding by waves breaking over the deck. Other navies took up the design and built their own breastwork monitors. One, HMVS *Cerberus* (the first to be built), was still in service with the Royal Australian Navy into the 1920s.

Discovery and Recovery

Several searches for the wreck of the *Monitor* in the 1940s and 1950s were inconclusive. She was finally found in 1973 and her identity confirmed the following year. She was lying in 220 feet (67 m) of water about 16 miles (26 km) south-southeast of Cape Hatteras Lighthouse. The wreck and the area around it were designated the first U.S. marine sanctuary, the Monitor National Marine Sanctuary, to prevent divers and salvage companies from disturbing her. The hull was found to be in such a poor state that recovering the whole vessel was impractical, but it was decided to recover her engine, propeller, guns and turret. When the turret was recovered in 2002, two skeletons were found inside it, but their identities could not be determined.

LEFT: *HMS* Glatton *is an example of a breastwork monitor. She was built at Chatham Dockyard and launched in 1871. She served with the Royal Navy until she was broken up in 1903.*

CUTTY SARK

The clipper ships of the 19th century were the most elegant and graceful ships of their time. They were the matinee idols of the maritime world. Built for speed, they had slender, streamlined hulls and carried a vast spread of sails. The ship that best epitomizes this beautiful type of vessel is the *Cutty Sark*. She was the fastest ship of her time, enabling her to dominate the lucrative wool trade between Australia and Britain for 10 years.

TYPE: composite clipper ship

LAUNCHED: Dumbarton, Scotland, 1869

LENGTH: 280 ft (85.4 m)

TONNAGE: 963 gross registered tons

CONSTRUCTION: East India teak and American rock elm on an iron frame

PROPULSION: 32 sails on three masts and bowsprit (1870: ship rig, 1916: barquentine rig)

The first ships to take the "clipper" name were the small, fast Baltimore clippers that plied their trade along America's Atlantic coast and in the Caribbean in the 18th century. They evolved into the bigger oceangoing clippers of the 19th century. The first of these was probably a ship called the *Ann McKim*, built in Baltimore in the early1830s. She had a longer hull than a Baltimore clipper and was square-rigged instead of the Baltimore clipper's fore-and-aft rig. She led to even bigger clippers such as *Rainbow* (1845) and *Sea Witch* (1846), which influenced the design of later ships. Clippers set record after record for speed. The *James Baines* set a transatlantic record of 12 days from Boston to Liverpool in September 1854 (a sailing record that still stands today). In the same year, *Lightning* set a one-day sailing record of 436 nautical miles (500 miles or 800 km) and *Flying Cloud* set a record of 89 days from New York to San Francisco that stood for 135 years.

Clippers satisfied the demand for fast ships to reach California during the gold rush. When gold was found in Australia, clippers carried prospectors there too. They were used in the opium trade between Britain, India and China, and served as blockade runners during the Civil War, but they really came into their own in the race to bring the first tea of the season back to England from China. The fastest ships commanded the highest prices for their cargoes and attracted the best captains and crews. Newspapers covered the races between them and gamblers placed bets on the outcome.

LEFT: *A clipper ship under full sail in a howling gale was an impressive sight. This painting by Antonio Jacobsen shows the American clipper* Flying Cloud *in 1913.*

ABOVE: Cutty Sark, *painted by Brazendale Cunnelly, dwarfs local junks off the coast of China. With all 32 sails set, she slices through the waves at top speed. Her complex sail plan required 11 miles (18 km) of rigging.*

From the Tea Run to the Wool Run

The clipper tea trade collapsed soon after the Suez Canal was opened in 1869. The canal shortened the journey from China by nearly 4,000 miles (more than 6,000 km). However, the winds on the new route through the canal didn't suit sailing ships, so the clippers continued to sail the long route around the Cape of Good Hope to take advantage of the trade winds.

Paradoxically, this was the very time when the most famous clipper was built. *Cutty Sark*, named after a character in Robert Burns's poem "Tam O'Shanter," was built with a timber hull over an iron frame. Her composite structure gave her hull the strength to carry more sail than other clippers so that she could be driven harder at a higher speed. Her top speed was 17.5 knots (20 mph or 32 km/h).

From her very first tea run to Shanghai and back in 1870 she found herself in competition with steamships. She made eight of these voyages to China, carrying wine, spirits and beer on the outward journey and returning with tea. After her last tea run in 1877, she carried a variety of cargoes between America, Japan, China, India and Australia. In 1880, there was such discontent and mutiny among the crew that the captain, James Wallace, jumped overboard into the shark-infested waters of the Java Sea and vanished. Her next captain,

SAILING SHIP OVERHAULED AND PASSED US!
From the log of the P&O steamship *Britannia*, whose crew were amazed to see *Cutty Sark* passing them while they were speeding along at a brisk 15 knots (17 mph or 28 km/h) on July 25, 1889

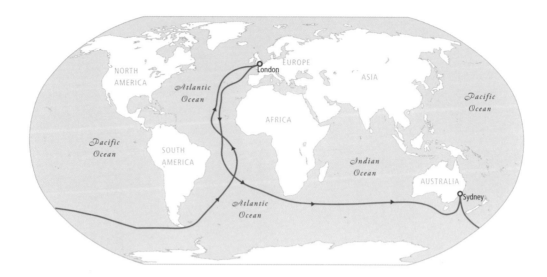

ABOVE: *The route used by clipper ships like* Cutty Sark *on the wool run from Australia to Britain took advantage of some of the windiest regions on Earth.*

William Bruce, was such an incompetent drunkard that he was removed from the ship and a new captain, F. Moore, was appointed.

By the 1880s, *Cutty Sark* had switched to the "wool run" from Australia. She sailed from Australia to Britain in only 83 days, 3 to 4 weeks faster than other ships. In 1885, Captain Moore was replaced by Richard Woodget. He pushed the ship to its fastest passage times by sailing farther south than other vessels to catch the strong westerly winds called the Roaring Forties. It was a risky tactic, because it took *Cutty Sark* into waters where icebergs might be encountered. Woodget managed to cut the Australia to Britain passage to only 73 days, enabling *Cutty Sark* to dominate the wool trade for the next 10 years. But it was not long before steamships took over this too. *Cutty Sark* was sold to a Portuguese company in 1895 and renamed *Ferreira*. In about 1916, she lost her masts in bad weather. Because of a shortage of materials during World War I, she was converted from square sails to a fore-and-aft barquentine rig.

LEFT: *By the 1920s, Cutty Sark's days as a fast cargo ship were over. Seen here anchored off Falmouth, she was opened to the public and used as a sail training ship.*

THERMOPYLAE

One of *Cutty Sark*'s major rivals was a clipper called *Thermopylae*. Built in 1868, she had set a record of 63 days from England to Melbourne, Australia, on her maiden voyage. In 1872 the two clippers raced against each other from Shanghai to London. They left Shanghai on June 17. On August 15, when *Cutty Sark* was 400 miles (640 km) in the lead, she lost her rudder in a storm off the Cape of Good Hope. Her crew did the impossible — they made and fitted a new rudder at sea in less than a week. They had to do it all again soon afterward when the replacement rudder broke free. *Thermopylae* took advantage of her misfortune and surged into the lead. *Cutty Sark* arrived in London on October 19, 7 days after *Thermopylae*. Whereas *Cutty Sark* has been preserved, *Thermopylae* ended her days as a naval training ship in Portugal before being sunk in 1907 by the Portuguese navy off Cascais.

BELOW: *A clipper called* Thermopylae *took part in a famous race against* Cutty Sark *from Shanghai to London in 1872.*

In 1922, she was sold to a new Portuguese owner, who renamed her *Maria do Amparo*. Later the same year, she was bought by Wilfred Dowman, who restored her original name and used her as a training ship in Cornwall. When he died in 1936, his widow gave the ship to the Thames Nautical Training College, which used her for training naval cadets. The journey from Falmouth, Cornwall, to London was the grand old lady's last sea voyage. Her working life finally ended in 1954, when she was put on permanent display in a custom-built dry dock at Greenwich, London.

Survival Against the Odds

It is something of a miracle that *Cutty Sark* still exists. On May 21, 2007, television viewers watched in horror as pictures showed the historic ship burning fiercely from end to end. It was later found that the fire had been started by an industrial vacuum cleaner that had been left running and overheated. It looked like a scene of total destruction. However, most of the ship's timbers had been removed for conservation and were not on site, and her iron frame was repairable. She was fully restored and reopened to the public by 2012. Two years later, a second fire damaged her deck, but it was detected and extinguished before it could spread. Against the odds, *Cutty Sark* has survived storms, icebergs, rot, mutiny and fires to become one of the most popular tourist attractions in London and a great ambassador for the age of the clipper ships.

FRAM

In the late 1800s and early 1900s, a series of expeditions set out to cross the last great unexplored wildernesses on Earth: the polar regions. One of the risks facing polar explorers was that their ship might become trapped in the sea ice. Once trapped, the ice pressure would crush the ship's hull. Then one designer created a ship that could not be crushed and, instead, took advantage of the ice.

TYPE: topsail schooner

LAUNCHED: Larvik, Norway, 1892

LENGTH: 127 ft 8 in (38.9 m)

TONNAGE: 402 gross registered tons

CONSTRUCTION: greenheart outer hull and pine inner hull on an oak frame

PROPULSION: sails on three masts and bowsprit plus a 220 hp (165 kW) triple-expansion steam engine (diesel engine from 1910), driving a single screw propeller

When the Norwegian explorer, Fridtjof Nansen (1861–1930), read reports of debris washing up on the coast of Greenland, it gave him an idea for a novel way to reach the North Pole — but it would require a ship the like of which had never been built before. That ship, *Fram* (*Forward*), took part in three landmark voyages of polar exploration.

The debris washed up on Greenland was from a ship called *Jeannette*. She had become trapped in pack ice in September 1879. She had drifted 600 miles (965 km) with the ice before she was crushed and sank near the New Siberian Islands. The distance she had traveled in the ice and the path of the debris suggested to Nansen that there was a transpolar ocean current that he might be able to use to reach the North Pole, or get close to it. If he deliberately sailed a ship into the pack ice at the right location, the current would carry the ice and the ship toward the pole. If his plan was to work, he would need a unique ship. The *Jeannette* had been strengthened for the polar ice, but it was still crushed. Nansen imagined a ship with a hull shaped so that the ice would push it upward instead of crushing it.

RIGHT: *Norwegian photographer Anders Beer Wilse took this photograph of the* Fram *in 1909 shortly before she departed for her third expedition, her first to the Antarctic.*

x

FIFTY SHIPS THAT CHANGED THE COURSE OF HISTORY

110

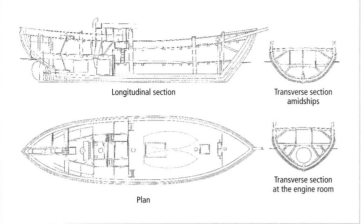

Longitudinal section

Transverse section amidships

Plan

Transverse section at the engine room

The ship, built by Scottish-Norwegian shipwright, Colin Archer, was not the most beautiful vessel. She had a blunt bow and stern, and her hull had a smooth, round cross-section. There were no projections that might catch on the ice. The rudder and propeller could be raised into wells in the bottom of the hull to protect them from damage. Although the ship was not expected to have to resist the full pressure of the ice around her, Archer made her incredibly strong, just in case. Her hull had four layers. Inside the oak ribs and frames, she was planked with 4-inch (10-cm) thick pitch pine. Outside the ribs and frames, she was covered with two layers of oak planking. These were then covered with a sheathing of greenheart, one of the densest woods available. The greenheart was fitted so that it could be torn away by ice without damaging the hull underneath. Iron straps at the bow and stern completed the structure. Inside, the living quarters were heavily insulated to keep them warm.

Entering the Ice

Nansen sailed for the Arctic in *Fram* on June 24, 1893, with enough food and supplies on board for 6 years. Her crew immediately discovered that her smooth, round hull caused her to roll violently. She collected her final supplies in Norway and 34 sled dogs in Siberia. She reached the pack ice on September 25, 1893, becoming the first ship ever to sail deliberately into the ice with

LEFT: *The* Fram *survived 3 years trapped in pack ice, traveled further north than any other ship, and then emerged from the ice into the open sea again unscathed.*

- - - - - - - Projected route
————— Actual route of crew
- - - - - - - Return route of *Fram*

North Pacific Ocean

BERING SEA

SEA OF OKHOTSK

ALASKA

SIBERIA

ARCTIC CIRCLE

CANADA

E. SIBERIAN SEA

BEAUFORT SEA

Arctic Ocean

North Pole

KARA SEA

Franz Josef Land

Spitz Bergen

BARENTS SEA

GREENLAND

GREENLAND SEA

NORWEGIAN SEA

NORWAY SWEDEN FINLAND

RUSSIA

ICELAND

North Atlantic Ocean

the intention of becoming trapped. The crew had no idea whether *Fram* would ever be released by the ice, because no one had ever done this before. Her engine, not needed any more, was dismantled and stored. A windmill 12 feet (3.6 m) across was put up to drive a generator to produce electricity for the lights. The dogs were taken off the ship to live on the ice, but had to be brought back on board after a polar bear ate two of them! The *Fram* did precisely what she was designed to do. As the ice tightened its viselike grip, the ship was pushed upward like a slippery round nut squeezed between finger and thumb.

As Nansen had predicted, the ship began moving north with the ice. After a year, she had traveled 189 miles (304 km). In March 1895, Nansen and Hjalmar Johansen decided to leave the ship and continue with the dog sleds, but they failed to reach the pole because of the constantly moving pack ice. *Fram* drifted in the ice for another 18 months, finally emerging on August 13, 1896 near West Svalbard, Norway. She was the first ship ever to be frozen into the pack ice and survive to sail away. Meanwhile, Nansen and Johansen spent the winter on the ice and were then picked up, by chance, by the British explorer F. G. Jackson, who took them to Norway.

In 1898, *Fram* was under way again. This time she had a false keel, added to improve her handling at sea. A new deck was added too, to

ABOVE: *With the* Fram *in the background, some of the crew venture out onto the ice. They took daily weather readings and calculated their position as the drifting ice carried the ship westward.*

RIGHT: *Musical entertainment helped to pass the time while the* Fram *was trapped in the ice. Here the ship's electrical engineer Bernhard Nordahl plays the organ (left), expedition doctor Henrik Greve Blessing sings (center) and explorer Fredrik Hjalmar Johansen plays the accordion (right).*

enlarge the living quarters. Her commander and expedition leader this time was Otto Sverdrup (1854–1930), who had taken part in the Nansen expedition and commanded *Fram* when Nansen left the ship. Sverdrup's expedition spent 4 years exploring fjords on Ellesmere Island and taking trips inland by dog sled, amassing a huge amount of scientific data.

The Final Expedition

In 1910, *Fram* was needed again. Her steam engine was replaced by a diesel engine, the first to be installed in a polar exploration ship. For her third expedition she was taken to the other end of the world for an

ABOVE: *In 1910, the Fram's steam engine was replaced by this 180 hp (135 kW) Swedish marine diesel engine for her third expedition. It was smaller, more powerful and easier to operate than the steam engine.*

attempt to reach the South Pole. The explorer on this occasion was Roald Amundsen (1872–1928). He had wanted to use *Fram* in the same way as Nansen for another attempt on the North Pole, but when he heard that Robert Peary (1856–1920) claimed to have got there already, he switched his attention to the South Pole.

He did not tell the ship's crew about the change until they were at sea! By keeping his plans secret, he managed to get to the Antarctic before the British explorer, Robert Falcon Scott (1868–1912), who was also planning an attempt on the South Pole. Amundsen arrived at Antarctica in January 1911 and set up camp to wait for the earliest opportunity to go for the pole the following spring. Amundsen and four colleagues reached the South Pole on December 14, 1911, 5 weeks before Scott, who died with his four companions during the journey back to their base camp. While Amundsen was on the ice, *Fram* explored Antarctic waters.

Fram sailed a total of 54,000 nautical miles (62,140 miles or 100,000 km) on her three expeditions, traveling farther north and farther south than any other ship. World War I put a stop to polar exploration, so *Fram* spent several years tied up at Horten, Norway. Her condition worsened during the 1920s until she was in danger of being taken to the breaker's yard. One of her old commanders, Otto Sverdrup, led a campaign to restore the historic ship. In May 1935, after Sverdrup's death, she was hauled out of the sea for the last time at Oslo and installed in her own purpose-built museum.

SPRAY

Although some sailors who went to sea for sport concentrated on short races, others pitted themselves against the elements on increasingly long voyages. The ultimate goal of these endurance sailors was a single-handed round-the-world voyage. Many people thought it was impossible. It was achieved for the first time in 1898, not in a purpose-built pedigree vessel, but in a converted fishing boat.

TYPE: gaff-rigged sloop, later re-rigged as a yawl

LAUNCHED: unknown

LENGTH: 39 ft 9 in (12 m)

TONNAGE: 12.7 tons

CONSTRUCTION: wooden carvel-planked hull

PROPULSION: sail

BELOW: *This photograph of the 55-year-old Joshua Slocum appeared in* Century Magazine *in September 1899, a year after the successful conclusion of his circumnavigation in the yacht,* Spray.

Joshua Slocum (1844–1909) had the sea in his blood from childhood. His family's home in Nova Scotia was within sight of the spectacular Bay of Fundy. After several failed attempts, he managed to run away to sea at the age of 14. He worked as a cabin boy and cook on a fishing boat. Two years later he signed on as an ordinary seaman on a merchant ship. By the age of 18 he had qualified as a second mate and quickly rose to the rank of chief mate.

Slocum proved to be a courageous and resourceful sailor. When one of his ships, *Washington*, was blown ashore in Alaska and started breaking up, he not only saved his wife and the crew, he managed to transfer most of the cargo to the ship's small boats and bring it ashore.

While he was stranded in the Philippines without a ship, he undertook some shipbuilding work in exchange for a 90-ton schooner called *Pato*, the first vessel he owned. He used it to start his own freight business, carrying cargo up and down the west coast of the United States and to Hawaii. It enabled him to buy more boats. One of them, *Aquidneck*, was wrecked in 1887 on the coast of Brazil. Undaunted, Slocum salvaged all the materials he could from the wreck and used them to build a new boat, *Liberdade*, which he sailed home. He published a book about the adventure, called *Voyage of the Liberdade*. He wrote a second book in 1893 about another eventful voyage to deliver a leaking steam-powered torpedo boat called *Destroyer* from New York to Brazil.

Rebuilding Spray

In 1892, a friend offered Slocum an old boat that he said needed some repairs. It turned out to be an oyster-fishing boat called *Spray*. She was in such a poor state that she had not been in the water for 7 years. She needed to be almost completely rebuilt. Slocum took her on and did the work himself. It took him more than a year to replace all the rotten timbers, install a new mast and make her

RIGHT: *Slocum's great voyage alone around the world made him famous. His route involved three Atlantic crossings and a series of adventures that he recounted in his best-selling book about the voyage.*

ABOVE: *From Slocum's book,* Sailing Alone Around the World, *this photograph shows* Spray *sailing in Australian waters, with Slocum at the helm.*

seaworthy. He decided to use *Spray* for an attempt on the round-the-world voyage.

He left Boston harbor on April 24, 1895, and sailed up the east coast to his boyhood home in Nova Scotia before setting out into the Atlantic on July 3. First, he crossed to Gibraltar, where he was attacked by pirates. On the advice of British naval officers, he turned around and sailed west instead of continuing east. He arrived at Rio de Janeiro on November 5. It was around this time that he shortened the boat's bowsprit and boom, and added a short mizzenmast and sail at the stern, transforming the boat from a sloop to a yawl. He sailed down the coast of South America to Cape Horn, where his sails were ripped by storms and he had to fire his gun to scare off some indigenous Patagonians who were trying to board *Spray*. He finally rounded Cape Horn on April 13, 1896.

Spray was so good at holding a course that Slocum was able to sail for long distances across the Pacific without touching the wheel. He navigated mostly by dead reckoning, using a tin clock he'd bought for one dollar. He stayed a while in Australia and then in South Africa before starting out across the Atlantic again on the homeward leg. He crossed his outward track across the ocean on May 8, 1898, and made a note of his circumnavigation in his log book. On June 27, 1898, he arrived at Newport, Rhode Island. He had sailed 46,000 miles (74,000 km).

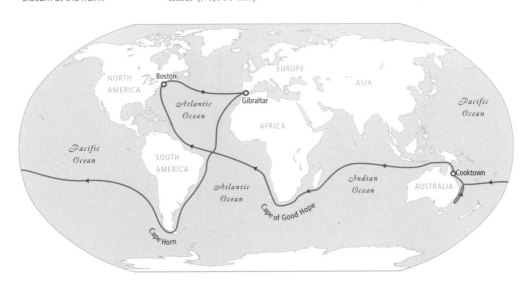

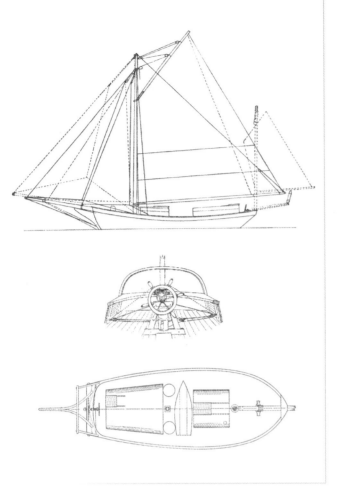

Slocum's account of his voyage, *Sailing Alone Around the World*, was published in 1899, making him world-famous. The proceeds of the book and the many talks he gave about his adventure enabled him to buy a small farm on Martha's Vineyard, Massachusetts, but he wasn't able to settle down. He went back to the sea with *Spray*. He sailed around the Caribbean during the winter and returned to New England each summer. By 1909, his income from book sales and talks was falling and he needed another money-spinning project. He was said to be considering an expedition to explore the Orinoco, Rio Negro and Amazon rivers.

ABOVE: *Solid lines show the Spray's sail-plan at the start of Slocum's voyage. Dotted lines show the yawl-rig that Slocum switched to in South American waters. His arrangement for lashing the wheel is shown in the middle.*

RIGHT: *Spray is seen here sporting a new set of sails given to Slocum by Commodore Mark Foy in Sydney, Australia, between October and December 1896.*

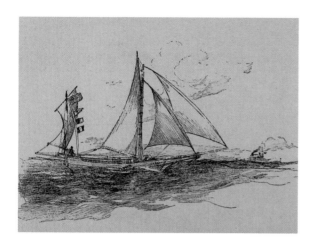

Fate and Legacy

Spray is undoubtedly a historical sailing vessel, but she isn't on display in a museum; she didn't rot away in a harbor somewhere; and she wasn't broken up. On November 14, 1909, the 65-year-old Slocum set sail from Vineyard Haven on Martha's Vineyard, bound for South America. Neither he nor *Spray* was ever seen again. Surprisingly, for someone who spent so much of his life on the sea, Slocum could not swim. He was declared legally dead in 1924. Some people think he was run down by a steamer. Others think *Spray* must have capsized. To this day, no one knows what happened to Joshua Slocum and *Spray*.

Slocum's rebuilt fishing boat was very influential. Many sailors wanted their own *Spray*, or at least a copy of it. Thousands of replicas have been built. Hundreds of sailors followed in Slocum's wake and made their own circumnavigations. His voyage also inspired the Vendée Globe and Around Alone yacht races.

GOING AROUND FASTER

Seventy-one years after Joshua Slocum set sail to circle the globe, a British sailor in his sixties set out to do the same thing, but faster. Where Slocum had taken 3 years, Francis Chichester wanted to follow the old clipper route and get around much faster. He left Plymouth on the south coast of England on August 27, 1966, in his yacht *Gipsy Moth IV*. (He named his yachts after the de Havilland Gipsy aircraft he flew in the 1930s.) He circled the world and returned to Plymouth just 274 days later (226 sailing days). A quarter of a million people on the coast and in thousands of small boats welcomed him home. He had made the fastest circumnavigation by any small vessel and the first ever true circumnavigation by the three great capes of Good Hope, Leeuwin and Horn, making only one stop on the way. He was knighted by Queen Elizabeth II using the same sword that Queen Elizabeth I had used to knight Francis Drake 400 years earlier. In 2005, the British yachtswoman Ellen MacArthur made a record-breaking solo nonstop circumnavigation in only 71 days 14 hours.

USS Oregon

At the end of the 19th century, the U.S. Navy, alarmed at the acquisition of modern European battleships by South American countries, started building its own new generation of battleships, the first since the Civil War. The USS *Oregon* was one of them. She earned her place in history by highlighting the need for a canal between the Pacific and Atlantic Oceans, thereby guaranteeing the completion of the bankrupt Panama Canal project, which might otherwise have been abandoned.

TYPE: *Indiana*-class coastal battleship

LAUNCHED: San Francisco, 1893

LENGTH: 350 ft 2 in (107 m)

TONNAGE: 10,288 long tons (10,453 metric tons) displacement

CONSTRUCTION: armored steel plate

PROPULSION: two vertical inverted triple-expansion reciprocating steam engines driving two screw propellers

In 1898, the United States was preparing for war against Spain. The Cuban War of Independence had been raging for 3 years. Cuban rebels had gained the upper hand and the war appeared to be drawing to a close. But when Spanish loyalists rioted in January 1898, the United States sent an armored cruiser, the USS *Maine*, to Cuba as a show of force to protect U.S. interests. On the evening of February 15, the *Maine* suddenly exploded and sank at her moorings in Havana harbor with the loss of 261 lives. Opinion in America blamed the sinking on a Spanish mine, and war with Spain seemed inevitable. America's North Atlantic Squadron needed reinforcement, so the *Oregon* was ordered to join it.

The *Oregon* had been launched in 1893 as a coastal defence battleship, a sister ship to the *Indiana* and *Massachusetts*. She was heavily armed and heavily armored for a ship of her size. Her main armament was a pair of twin 13-inch (33-cm) guns mounted in turrets on her centerline and four twin 8-inch (20-cm) guns, also in turrets. She had

RIGHT: *The USS* Maine *rests on the bottom of Havana harbor after a mystery explosion — the event that prompted the* Oregon's *high-speed transit from the Pacific to the Atlantic.*

RIGHT: *The USS* Oregon *(BB-3), seen here at anchor near San Francisco in 1916/1917 during her time in the reserve fleet, was nicknamed the Bulldog of the Navy.*

ARMOR

The first armored warships were clad in ordinary iron plate, often with a thick wooden backing. In the early 1880s, compound armor was developed. It was composed of high-carbon steel laid over wrought iron. The hard steel plate fragmented incoming missiles and the softer back-plate caught the splinters. At the end of the 1880s, multilayer compound armor was replaced by single-layer nickel-steel armor. The USS Oregon was fitted with a type of armor described as Harveyized, or Harvey, armor. Developed in 1890, it was a single steel plate whose front face had been cooked at high temperature with charcoal to harden it by increasing its carbon content. By the end of the century, Harvey armor was replaced by Krupp armor, made of steel with added chromium and a front face hardened by using carbon-rich gas.

up to 28 other guns of all sizes plus four torpedo tubes. Her armor was up to 18 inches (46 cm) thick.

She was commissioned in 1896 as America's first battleship on its Pacific coast, but now she had to get to the Atlantic. She had to make a journey of 15,700 miles (25,300 km) around South America. She was delayed at Cape Horn by a severe storm and had to make several stops on the way to replenish her coal supplies. The journey took just over 2 months.

The Need for a Canal

Coverage of the *Oregon's* voyage in the United States had a dramatic effect. Although the government and public congratulated her crew on their 66-day voyage from coast to coast, it was immediately obvious that a delay of more than 2 months to transfer warships from one ocean to the other was unacceptable in wartime. A canal linking the two oceans was urgently needed. Fortunately, one was already under construction; but the project was facing severe difficulties.

USS OREGON

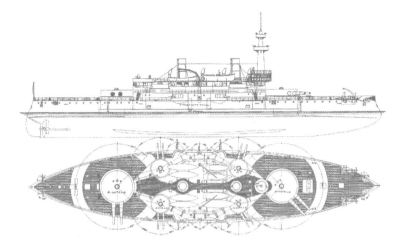

THE *INDIANA* CLASS

In 1889, the U.S. Secretary of the Navy proposed building dozens of new battleships and over 150 other warships. It appeared to signal the end of the isolationist stance America had held since the start of the 19th century. This was expressed in the Monroe Doctrine of 1823, named after President James Monroe. It stated that America would not meddle in the affairs of European countries and would resist any attempt by European countries to colonize states in North or South America. The United States also declined to form alliances with the European powers from which many of its citizens and their forefathers had fled. As a result, the proposed naval construction program of 1889 was voted down in Congress, which instead approved the construction of just three small *Indiana*-class coastal battleships, one of which (the *Oregon*) had to be built on the west coast. They were called "coastal battleships" to reassure Americans that these ships were not for foreign military adventures, but were intended for defense.

By the mid-1890s, the French attempt to build the Panama Canal was bankrupt and completion of the project looked unlikely. Now keener than ever for a central American canal, the United States stepped in and acquired the project. Panama, then part of Colombia, was encouraged and assisted by America to declare its independence. The new Republic of Panama then leased the canal zone to the United States. It took another 11 years to complete the canal, which opened on August 15, 1914.

A New Emergency in the Far East

When the *Oregon* reached Cuba, she joined the blockade of Santiago de Cuba, trapping the Spanish fleet in the port. When several of the American warships had to leave to take on coal, the Spanish ships took the opportunity to break out of the port. The ensuing Battle of Santiago de Cuba, in which the *Oregon* played an important part, was a resounding victory for the American ships. Spanish resistance to Cuban independence melted away and the conflict ended soon after.

The *Oregon* was ordered to New York for a refit and then returned to the Pacific. As a result of the wider Spanish–American War that had begun with the hostilities in Cuba, the United States acquired several Spanish territories, including the Philippines.

However, the Philippines had declared independence, which the United States did not recognize. The two countries were soon at war and the *Oregon* was ordered into action again.

She spent the next few years in Philippine, Japanese and Chinese waters. She was badly damaged when she ran aground near the Changshan Islands off the Chinese coast and needed repairs that took more than a year. In 1906, she returned to the United States for an extensive refit and was then held in reserve for much of the next 11 years.

Although the *Oregon* was preserved for a time as a floating monument and museum, at the outbreak of World War II the government decided to break her up for her valuable and much-needed steel, which was in short supply. She had a brief reprieve during the war when, stripped of her guns and superstructure, she was used as an ammunition barge at Guam, where she stayed after the war. Then, in November 1948, she broke loose from her moorings during a typhoon and drifted out to sea. She was found 3 weeks later and towed back to Guam, but her days were numbered. Her teak decks and steel armor plate were removed and then she was towed to Japan and scrapped in 1956.

USS HOLLAND

The first successful submersible vehicle was built in 1620 by Dutchman Cornelius Drebbel and demonstrated in the River Thames in London. After that, there were numerous attempts to construct practical naval submarines to carry out underwater attacks on surface ships. Some had limited success, but most were failures until Irish inventor John Philip Holland finally cracked the problems in the late 19th century. His *Holland VI* submarine was bought by the U.S. Navy and became its first modern naval submarine in 1900.

TYPE: submarine

LAUNCHED: Elizabeth, New Jersey, 1897

LENGTH: 53 ft 10 in (16.4 m)

TONNAGE: 64 long tons (65 metric tons) displacement surfaced, 74 long tons (75 metric tons) submerged

CONSTRUCTION: steel plate on iron frame

PROPULSION: 1 x Otto gasoline engine, 45 bhp (34 kW), 1 x Electro Dynamic electric motor, 75 bhp (56 kW), 66-cell Exide battery, single screw propeller

John Philip Holland (1841–1914) emigrated from Ireland to the United States in 1873 at the age of 32. He worked as a teacher in Paterson, New Jersey, but his real passion was submarines. His attempts to interest navies in building submarines fell on deaf ears, but Irish expatriates in the United States were prepared to fund his work. They were interested in using submarines to attack British ships as part of their campaign for Irish independence from Britain. Between 1878 and 1883, Holland designed and built three submarines, but they met with limited success and his backers eventually withdrew their support. Holland, who had quit his job as a teacher to work full-time on his submarines, went to work for the Pneumatic Gun Company, which funded the construction of a fourth submarine. This underwent sea trials but was not pursued any further.

RIGHT: *John P. Holland emerges from the conning tower hatch of his* Holland VI *submarine, later to become the USS* Holland *(SS-1).*

Then the U.S. Navy announced a design competition for a submarine torpedo boat. It had to be capable of 15 knots (17 mph or 28 km/h) on the surface and just over half that speed submerged, with two hours' underwater endurance, torpedo armament and a dive depth of 150 feet (45 m). Holland entered the competition and won with his next submarine design, the *Holland V* or *Plunger*. After many delays over several years, the funding was approved at last and the *Plunger* could finally be built. Instead of the hand-cranked propellers of earlier submarines, the *Plunger* was powered by a steam engine on the surface and an electric motor when submerged. However, Holland found that the steam engine was impractical, as it made the submarine's interior unbearably hot.

ABOVE: *One of Holland's submarines is seen here while under construction. The large opening in the bow is the vessel's single torpedo tube.*

Success at Last

Holland quickly developed an improved design, the *Holland VI*. It was launched on May 17, 1897. Five months later, it suffered a near disaster when a workman left a valve open. Water flooded in and sank the submarine at the dockside. It took 18 hours to refloat it and several days to dry out its electrical equipment, but it survived and began sea trials in March 1898. A gasoline engine powered it on the surface and also charged its electric battery. When it dived, the gasoline engine was turned off and replaced by an electric motor powered by the battery that had been charged by the engine. It was an elegant solution. This gave the *Holland VI* a top speed of 9 mph (15 km/h) on the surface and 6 mph (9 km/h) submerged. After successful sea trials, the *Holland VI* was finally bought by the U.S. Navy for $160,000 (the equivalent of about 4.6 million dollars today), even though it had not met their minimum requirements. It entered service as U.S. Navy submarine USS *Holland* (*SS-1*) on October 12, 1900.

BUSHNELL'S *TURTLE*

In 1775, David Bushnell (1754–1824) built the first diving craft known to have been used in combat. It was an egg-shaped wooden vessel called the *Turtle* (see page 86). Just big enough for one person, it submerged by taking in water as ballast. Propulsion and depth were controlled by two hand-cranked screw propellers. On September 6, 1776, during the Revolutionary War, Sergeant Ezra Lee steered the *Turtle* toward a British ship, HMS *Eagle*, anchored in New York harbor. Lee tried to attach an explosive charge to the ship's hull, but failed.

Armed for Combat

The USS *Holland* was no toy or experimental craft. She was designed as a practical submersible warship. Her armament comprised a single 18-inch (45-cm) torpedo tube and an 8-inch (20-cm) "dynamite gun" that could lob explosive charges into the air. One torpedo was carried in the tube, with another two carried internally. The dynamite gun could hurl a 200-pound (90-kg) projectile called an "aerial torpedo" more than 1,000 yards (900 m). Originally, the submarine had two of these guns, but the aft gun was later removed.

The *Holland* carried a crew of six or seven. They initiated a dive by opening valves to let seawater flood into ballast tanks. Then horizontal vanes called diving rudders (hydroplanes today) were tilted to tip the sub's nose down. To return to the surface, the diving rudders were reversed and water was blown out of the ballast tanks by compressed air. One of the *Holland*'s shortcomings was poor visibility. In order to mount an attack, the sub had to come to the surface so that a crewman could look out through windows in the conning tower to line up the sub with its target. This gave away the submarine's presence and location, robbing it of the advantage of surprise. Periscopes were

THE *HUNLEY*

The *H.L. Hunley* was the first submarine to sink a ship. Named after her inventor, the CSS *Hunley* was 40 feet (12 m) long and operated by a crew of eight: seven to hand-crank the propeller and one to steer. She sank twice during trials, killing 13 crew members including Horace Hunley himself. On February 17, 1864, during the Civil War, the *Hunley* attacked the Union warship USS *Housatonic* near Charleston. She rammed a spear into the *Housatonic*'s hull, then backed away and detonated an explosive charge attached to the spear. The *Housatonic* sank, but the *Hunley* and her crew disappeared too. She was rediscovered on the seabed in 1995 and raised to the surface in 2000.

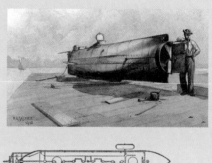

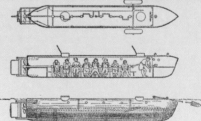

ABOVE: *The* Hunley *submarine was a simple iron tube with very little room inside for the crew. She was difficult and dangerous to operate, but proved surprisingly effective in combat.*

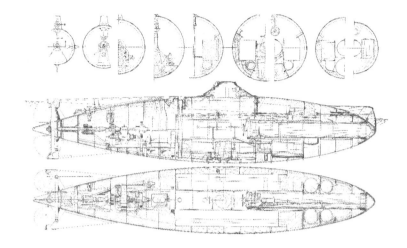

RIGHT: *Once the U.S. Navy had commissioned its first submarine, other navies became interested in Holland's submarines. The British Royal Navy commissioned Holland-1, the first of its new Holland class.*

already in use on land and they had been built into other submarines, but they were nonretractable. The retractable submarine periscope would be invented very soon, in 1902.

The USS *Holland* never saw action. She was used mainly for training the crews that would man later submarines. She also served as the prototype for the seven A-class submarines that followed her. She was decommissioned on July 17, 1905 and struck from the navy register on November 21, 1910. She was put on display in a New Jersey park until 1932, when she was scrapped.

Following the success of the USS *Holland*, the navies of Britain, Russia, the Netherlands and Japan all ordered Holland submarines, prompting other nations, notably Germany, to develop their own designs. Many of the formidable naval submarines in service today can trace their heritage back to the *Holland VI*.

RIGHT: *The Imperial Japanese Navy joined the growing Holland club when it bought five Holland VII submarines from the United States. They were a bigger and more powerful version of the Holland VI.*

USS HOLLAND

POTEMKIN

The effects of the Russian Revolution of 1917 spread far beyond the borders of Russia. The revolution swept away the last Czar and led to the establishment of the Soviet Union, with global implications. One of the first signs that revolution was a real danger to the old Czarist empire was a mutiny on a Russian warship, _Potemkin_, 12 years earlier. It encouraged the spread of revolutionary fervor and its failure, and the failure of rebellions elsewhere in Russia, enabled revolutionaries like Lenin to work out what they had to do to win next time, in 1917.

TYPE: pre-dreadnought battleship

LAUNCHED: Nikolayev Shipyard (now Mykolaiv), Ukraine, 1900

LENGTH: 378 ft 5 in (115.3 m)

TONNAGE: 12,900 long tons (13,107 metric tons) displacement

CONSTRUCTION: armored steel hull

PROPULSION: two vertical triple-expansion steam engines driving twin screws

The _Knyaz' Potemkin Tavricheskiy_ (_Prince Potemkin of Tauris_), better known as the battleship _Potemkin_, was built in the late 1890s for Czarist Russia's Black Sea Fleet. She was built to match the best battleships of other major naval powers. She was armed with 40 guns, including two twin 12-inch (305-mm) guns, and protected by the latest Krupp armor.

In 1904, war broke out between Russia and Japan over Russian expansion into the Japanese sphere of influence in the Pacific. Russia suffered a series of defeats by the Japanese navy, notably a devastating destruction of its Baltic Fleet at the Battle of Tsushima in 1905. Consequently, morale in the rest of the navy and more widely in the Russian population was very low. Revolutionaries among the crews of the Black Sea Fleet were planning a fleet-wide mutiny to inspire Russian peasants to rise up against the aristocracy. The mutiny was to begin in August, but events on board the _Potemkin_ intervened.

On June 27, 1905, just a month after the Battle of Tsushima, the battleship _Potemkin_ was preparing for gunnery practice off the Ukrainian coast when her crew was served a meal of borscht made with rotten meat riddled with maggots. The ship's doctor declared the meat fit for consumption, but the crew refused to eat it. The senior officers were outraged at this insubordination and the captain

LEFT: _Some of the battleship Potemkin's crew pose for a photograph shortly before the notorious mutiny. The lieutenant in the middle of the picture was one of the officers killed by the mutineers._

FIFTY SHIPS THAT CHANGED THE COURSE OF HISTORY

RIGHT: *This painting by Pyotr Timofeyevich Fomin captures the moment the Potemkin's crew took control of the ship by force from her officers.*

threatened to have the men shot. He appeared to be serious, because he summoned a squad of armed marines. It proved to be the last straw. Gunfire broke out between the officers and crew. A seaman called Grigory Vakulinchuk who appeared to be the crew's ringleader was shot by the ship's second-in-command, Ippolit Gilyarovsky, who in turn was seized by the men and thrown overboard. The marines failed to come to the officers' aid.

The crew took control of the ship. The remaining officers were locked in their cabins and the crew elected a committee to run the ship. Then they issued a manifesto saying:

To all civilized citizens and to the working people! The crimes of the autocratic government have exhausted all patience. The whole of Russia, burning with indignation, exclaims: "Down with the chains of bondage!" The government wants to drown the country in blood, forgetting that the troops consist of sons of the oppressed people. The crew of the Potemkin has taken the first decisive step. We refuse to go on acting as the people's hangman. Our slogan is: "Freedom for the whole Russian people or death!" We demand an end to the war and the immediate convocation of a constituent assembly on the basis of universal suffrage. That is the aim for which we shall fight to the end: victory or death! All free men, all workers will be on our side in the struggle for liberty and peace. Down with the autocracy! Long live the constituent assembly!

MUTINY ON THE *BOUNTY*

One of the most famous, or notorious, mutinies in naval history was the mutiny on HMS *Bounty* in 1789. The *Bounty* was a small British naval vessel sent to the Pacific to transport breadfruit plants from Tahiti to the West Indies to provide cheap food for slaves. Three weeks after the ship left Tahiti with her cargo of plants, master's mate Fletcher Christian led a mutiny involving almost half the crew. The ship's captain, William Bligh, and most of the loyal crewmen were cast adrift in *Bounty*'s launch. In an extraordinary feat of seamanship, Bligh navigated 4,170 miles (6,710 km) across the Pacific in the open boat to Timor in Indonesia. Meanwhile, the mutineers sailed to Pitcairn Island in the southern Pacific, where they burned the *Bounty*. By 1808, only one of the mutineers was still alive; the others had died from illness or were killed in fighting. When breadfruit plants were finally delivered to the West Indies, the slaves refused to eat the fruit!

BELOW: *The Bounty II, seen here on Lake Michigan in 2010, is a replica of HMS* Bounty, *the ship made famous by the 1789 mutiny led by the master's mate, Fletcher Christian.*

Under the Red Flag

Flying the red flag of revolution, they steamed to the Ukrainian port of Odessa, where rebellion was in the air and rioters were on the streets. There was such serious disorder in the city that martial law was declared and troops opened fire on the rioters. Up to 2,000 people were killed and another 3,000 were wounded. Revolutionaries in the city hoped for support from the *Potemkin* and its heavy guns, but the crew declined to get involved. With the army firmly in control of Odessa again, the *Potemkin*'s crew requested an amnesty. When it was refused, they put to sea hoping that other ships would join them. The crew of one ship, the battleship *St. George*, mutinied briefly until officers and loyal Czarist crewmen regained control.

The *Potemkin* headed for Constanta on the Romanian coast. Russian warships of the Baltic Fleet that had been ordered to stop her refused to open fire on her. When Romania insisted that the crew surrender the ship, she steamed on to Feodosia in the Crimea. Unable to secure supplies there, the crew finally had to admit defeat. They returned to Constanta and surrendered. As they left the ship, they opened the *Potemkin*'s sea-cocks and sank her in the harbor.

> ALL OF RUSSIA IS WAITING TO RISE AND THROW OFF THE CHAINS OF SLAVERY.
> **Afanasy Matyushenko, one of the leaders of the *Potemkin* mutiny**

Eisenstein's film,
Battleship Potemkin, which
told the story of the mutiny,
was heavily promoted using
stirring posters like this. The
Soviet Union saw the mutiny
as proof that troops would
sometimes join the people in
overthrowing the old order.

The *Potemkin* was refloated and, suffering from saltwater damage internally, towed to Sevastopol, where she was repaired and renamed *Panteleimon* after a Russian saint. During World War I, she took part in several actions against Turkish ships and coastal installations. After the Revolution in 1917, she was again named *Potemkin-Tavricheskiy.* A few months later her name was changed again, to *Borets za Svobodu* (*Freedom Fighter*).

BELOW: *The battleship*
Potemkin lies at anchor. She
appears to be flying a British
Union flag from her bow,
perhaps indicating that this
photograph dates from the
end of World War I when she
was seized by the Allies.

She was captured by German forces in May 1918 and handed over to the Allies at the end of the war. They left her in Sevastopol but disabled her engines so that she could not be used by approaching Bolsheviks. But both the White Russians and the Bolsheviks managed to use her at one time or another until she was abandoned in 1920 and finally scrapped in 1923.

The *Potemkin* and the 1905 mutiny of her crew were immortalized in the famous silent film made in 1925 by Sergei Eisenstein, *The Battleship Potemkin.* Most of the real *Potemkin*'s crew stayed in Romania after they left the ship. Some went to Argentina. Others who returned to Russia were arrested and executed. The last survivor, Ivan Beshof, found his way to Dublin, Ireland, where he died in 1987 at the age of 102.

HMS *DREADNOUGHT*

A new type of battleship dominated navies in the early 1900s: the dreadnought. It resulted from a dramatic change in naval warfare tactics and set off an arms race in battleship construction and armament between the world's leading navies.

TYPE: dreadnought battleship

LAUNCHED: HM Dockyard, Portsmouth, England, 1906

LENGTH: 527 ft (160.6 m)

TONNAGE: 18,120 long tons (18,410 metric tons) displacement

CONSTRUCTION: armored steel

PROPULSION: two pairs of direct-drive turbines driving four screw propellers

*B*attleship design underwent a revolution in the early 1900s. Torpedoes had become a serious danger to warships. They were more than capable of hitting ships, and sinking them, over their typical battle separation of about 3,000 yards (2.7 km). All the largest navies were thinking about fighting over longer ranges with bigger guns, but the first person to air the idea publicly was an Italian naval engineer, Vittorio Cuniberti. He wrote an article in 1903 proposing an "all-big-gun" battleship. Just one size of gun was needed, because fighting at long range rendered most of the smaller guns carried by existing battleships unnecessary. Cuniberti's ideal future battleship would be armed only with the biggest guns available. The usual procedure was to design a ship first and then fill it with guns. From now on, the selection of the guns would come first and then the ship would be designed around them.

The first all-big-gun battleship to be launched was the British Royal Navy's *Dreadnought*. She was armed with 10 12-inch (305-mm) guns in five twin-gun turrets. Each of these giant guns could hurl a shell weighing 850 pounds (390 kg) a distance of more than 10 miles (16 km). *Dreadnought* was also the first battleship to be powered by steam-turbine engines, giving the massive vessel a top speed of 21 knots (24 mph or 40 km/h) — faster than any other battleship afloat.

RIGHT: *The quarterdeck (stern area) of HMS* Dreadnought *is shown here in 1910, cleared for action. Her two aft twin-gun turrets are clearly visible, with a pair of 12-pounders mounted on top of the turret in the foreground.*

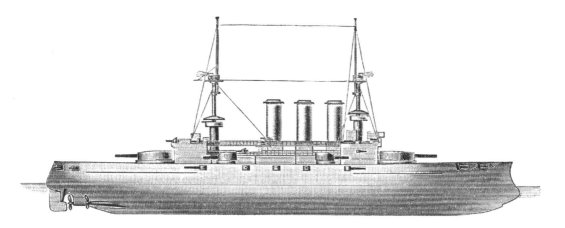

ABOVE: *This is the ship that the Italian designer Vittorio Cuniberti envisaged as the ideal all-big-gun battleship for the Royal Navy. It was developed into the design that became the HMS* Dreadnought.

HMS *Dreadnought* was intended to act as a deterrent to any nation thinking of attacking Britain. She was such a fast and powerful fighting vessel that she immediately rendered every other battleship obsolete. But other navies had been thinking along the same lines and soon built their own dreadnoughts. Japan had actually started building its first dreadnought, the *Satsuma*, before Britain, but *Dreadnought* was launched first. America's first dreadnought, USS *Michigan*, followed in 1908. The United States had been prompted to embark on a new warship construction program by the emergence of Japan as a serious naval power in the Pacific. Meanwhile in Europe, Britain was increasingly alarmed by the number of warships being built by Germany; they represented the first serious challenge to Britain's naval supremacy since Nelson's time. The result was a worldwide explosion in battleship construction, with each major naval power watching what the others did and then matching or surpassing it.

HMS *Dreadnought*'s technological lead did not last long. The first dreadnoughts were followed by even bigger and more heavily armed ships known as superdreadnoughts. The British were first again, with their *Orion*-class ships, but other nations quickly followed. They mounted bigger and bigger guns, ultimately 15-inch (380-mm) weapons. During this time there was also a change of fuel, from coal to oil. Oil packed more energy into a smaller volume, so oil-fired boilers could be smaller.

THE STEAM TURBINE

Until HMS *Dreadnought*, steam-powered battleships had big, heavy, inefficient reciprocating engines — engines worked by pistons flying back and forth. Heavy mechanical linkages converted the pistons' straight-line movements into rotary motion to turn the ship's propellers. Charles Parsons's steam turbine was simpler, smaller and lighter. HMS *Dreadnought*'s turbines saved almost 1,000 tons (more than 1,000 metric tons) in machinery weight. Turbines were also more efficient than piston engines, because they converted steam pressure directly into rotary motion. As well as making warships faster than ever, turbine engines vibrated less than reciprocating engines and needed less maintenance.

Although she had been built for combat with other surface ships, the only action HMS *Dreadnought* saw during World War I was with a submarine. The German submarine *U-29* surfaced in front of her in the Pentland Firth, north of Scotland, on March 18, 1915. *Dreadnought* rammed the submarine and sank it with all hands.

The Battle of Jutland

Dreadnought battleships met in combat only once, at the Battle of Jutland during World War I. Ironically, HMS *Dreadnought* herself did not take part. The battle was fought between the Royal Navy's Grand Fleet, commanded by Admiral Sir John Jellicoe, and the German Navy's High Seas Fleet, commanded by Admiral Reinhard Scheer. The Royal Navy was blockading the North Sea to starve Germany of essential supplies and also to prevent the German navy from breaking out into the Atlantic where it could attack British merchant shipping. At the end of May 1916, a group of German battlecruisers ventured out into the North Sea to lure British ships out where the German fleet would be waiting for them. The German navy expected to be fighting only a small number of British ships. However, the British had learned that 40 German warships had left port and so they mobilized the entire Grand Fleet.

On the afternoon of May 31, a British force of 151 ships including 28 battleships met a German force of 99 ships with 16 battleships.

AN UNINVITED GUEST

In 1894, Charles Parsons (1854–1931), inventor of the steam turbine, built an experimental turbine-powered boat called *Turbinia* to show what his new steam-turbine engine could do. *Turbinia* had a

top speed of 39 mph (63 km/h), twice as fast as the fastest battleship. He demonstrated it in a very bold way. *Turbinia* arrived unannounced at Spithead on the south coast of England during the Navy Review for Queen Victoria's Diamond Jubilee in 1897. She sped up and down between the lumbering warships, and no naval vessel present could catch her. The navy was so impressed that it built two destroyers with steam-turbine engines, HMS *Viper* and *Cobra*, in 1899. They were so successful that the first turbine-powered battleship, HMS *Dreadnought*, followed in 1906, and the navy decided that all of its warships would have steam-turbine engines from then on.

The German ships scored first, sinking three British ships. The British had more success in the engagements that followed. The fighting went on into the night until, under cover of darkness, the German ships returned to port. The Royal Navy had lost 14 ships and more than 6,000 dead. Germany lost 11 ships and more than 2,500 dead. Both sides claimed victory. The British had lost more ships and suffered higher casualties, but they retained control of the North Sea and stopped the German fleet from breaking out.

After World War I, Germany was prevented from building new warships by the Treaty of Versailles. Britain, impoverished by the war, could not afford a new warship construction program and looked likely to be overtaken by other countries. However, none of the other major naval powers relished the vast expense of building new fleets. Consequently, the Washington Naval Treaty, signed in 1922 by the United States, Britain, Japan, France and Italy, limited the numbers, types and sizes of warships that could be built. In addition, the treaty required most of the old dreadnought-type ships to be scrapped. HMS *Dreadnought* herself had already been sold for scrap the previous year.

RMS *Lusitania*

The *Lusitania* was one of the biggest, fastest and most luxurious ocean liners of her day. Her deliberate sinking with enormous loss of life during World War I contributed to the United States' decision to enter the war against Germany.

TYPE: ocean liner

LAUNCHED: John Brown Shipyard, River Clyde, Scotland, 1906

LENGTH: 787 ft (240 m)

TONNAGE: 44,060 long tons (44,767 metric tons) displacement

CONSTRUCTION: riveted steel

PROPULSION: four direct-acting Parsons steam turbines, 76,000 hp (57 MW), driving four screw propellers

In the early 1900s, there was keen competition between Europe's biggest shipping lines operating liners on the lucrative transatlantic passenger route. One of them, Cunard, set out to build the fastest and most luxurious liners. Two new ships, RMS *Mauretania* and *Lusitania*, were launched in 1906. Both were fitted with steam-turbine engines three times more powerful than HMS *Dreadnought*'s, giving them a sustained cruising speed of nearly 30 mph (50 km/h), faster than any other liner.

On the morning of May 7, 1915, *Lusitania* was sailing eastbound along the south coast of Ireland, her 1,200 passengers unaware that she had received radio messages warning of U-boats in the area. A zigzag course was recommended in hostile waters, but *Lusitania* was steaming in a straight line.

On the eve of *Lusitania*'s departure from New York, submarine *U-20* had left Emden on the north coast of Germany. She headed around Scotland and down Ireland's Atlantic coast, aiming to enter the Irish Sea from the south to hunt for ships off the English coast at Liverpool, *Lusitania*'s destination. Foggy weather had let several ships

escape *U-20*. She had managed to sink a couple of small freighters, but her commander, Walther Schwieger, was impatient for bigger prey.

At 1.20 p.m., while *U-20* was on the surface charging its batteries, Schwieger spotted smoke in the distance. It was rising from four funnels, so this had to be a large ship, just the sort of target he'd hoped for, and it was steaming straight toward him. *Lusitania* was flying no flags, her distinctive red funnels had been painted black and her name had been painted out, but one of *U-20*'s officers correctly identified her as either *Mauretania* or her sister ship, *Lusitania*. *U-20* dived and waited. As soon as *Lusitania* came within torpedo range, Schwieger fired.

At least two of *Lusitania*'s crew spotted the torpedo and tried to sound a warning, but it was too late. The torpedo slammed into the ship's side and exploded. Almost immediately, a more powerful explosion threw up a mountain of water and debris. Despite the known risk to the ship, there had been no lifeboat drills, so passengers were unsure what they should do or even how to put their life jackets on. *Lusitania* turned to port in the hope of reaching the Irish coast and beaching before she sank. She was listing to starboard and sinking by the bow. Then her electrical power failed. The cabins and the maze of passages in the depths of the ship were suddenly pitch-black and passengers who had been coming up in the ship's elevators were instantly trapped.

The end came very quickly. A ship like *Lusitania* would normally be expected to stay afloat for several hours, even if badly damaged,

ABOVE: *The first-class dining saloon was the grandest room afloat. In the style of Louis XVI, it had white plaster, gold leaf, mahogany panels and Corinthian pillars, topped with a dome decorated with frescoes.*

RIGHT: *Fate brought the* Lusitania *and* U-20 *together with catastrophic results. The submarine's commander was keen to bag a large enemy target when he found the* Lusitania *steaming straight toward him.*

LEFT: Lusitania *arrives in New York in 1907 at the end of her maiden voyage. Spectators on the dockside stand on barrels to get a better view of the great ship.*

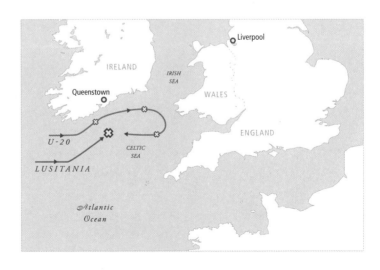

IN THE HISTORY OF WARS THERE IS NO SINGLE DEED COMPARABLE IN ITS INHUMANITY AND ITS HORROR.

The *New York Times*'s response to the *Lusitania*'s sinking

MAURETANIA

Lusitania's sister ship, *Mauretania*, enjoyed a long career. Launched 3 months after *Lusitania*, she quickly captured the Blue Riband for the fastest eastbound and westbound transatlantic voyages, previously held by the *Lusitania*. She held both speed records for nearly 20 years. During World War I, she served as a troop transport and hospital ship. She resumed her transatlantic passenger service after the war, eventually making 269 return crossings. She lost both of her speed records to the German liner *Bremen* in 1929. The following year, by then uncompetitive as a liner, she was reborn as a cruise ship. In 1934, Britain's two great rival liner operators, Cunard and White Star, merged and retired some of their older ships, including *Mauretania*. She was scrapped in 1935.

allowing time for other ships to come to her aid; but *Lusitania* sank in only 18 minutes. By the time rescue boats had made the short trip from the coast to the *Lusitania*'s last position, many of those in the cold water were already dead. A total of 1,195 passengers and crew lost their lives; just over 760 survived.

The scene at Queenstown harbor (now Cobh) was ghastly as scores of bodies were laid side-by-side at the water's edge. The town hall was used as a morgue. Survivors were faced with the awful task of walking along the lines of bodies trying to identify relatives and friends.

The dead included 123 American citizens. The reaction in the United States was predictable. There was outrage that Germany had, without warning, attacked and sunk an unarmed civilian vessel carrying citizens of a neutral nation. The U.S. ambassador in London urged his country to declare war on Germany.

Investigating the Cause

The sinking raised a number of questions. Why was *Lusitania* sailing so close to shore, where U-boats might be lying in wait? Why wasn't she zigzagging? There were suggestions that she had been placed deliberately in a U-boat hunting ground in the hope that she would be attacked and thus bring the United States into the war. What caused the second explosion? Was she carrying munitions, as Germany had claimed? Germany refused to apologize for the sinking, but changed the navy's rules of engagement so that passenger liners would not be attacked in future.

A Board of Trade inquiry found that the torpedo fired by the U-boat was the sole cause of the sinking. No evidence was ever uncovered to indicate that the ship had been maneuvered into a position where she was more likely to be attacked. And the captain's decision not to zigzag the ship farther from the Irish coast was explained by his determination to reach Liverpool on time without subjecting the passengers — some of whom were very wealthy — to a rolling, zigzagging course. But the cause of the second explosion was not determined. There were persistent allegations that the ship was carrying munitions, but her owners denied that she was carrying anything other than the 4 million rounds of rifle cartridges that were openly listed on her cargo manifest. A boiler explosion could explain the second blast, but there were no reports from the crew of a boiler exploding.

In 1993, Dr Robert Ballard, who had previously discovered the wreck of the *Titanic*, sent submersibles down to survey the wreck. His photographs show the ship lying on its starboard side, most of the superstructure rusted away, her hull distorted and collapsed. Significantly, Ballard found coal strewn about the seabed. It appeared to have fallen from holes in the hull after the explosions. *Lusitania* had a voracious appetite for coal. More than 5,000 tons had been loaded in New York and most of it had been burned during the voyage. It is thought that the torpedo explosion shook coal dust into the air inside the near-empty bunkers. The mixture of coal dust and oxygen was explosive and needed only a spark from somewhere to set it off, ripping the hull apart and sending the ship to the bottom in minutes.

RMS *Titanic*

In 1912, the biggest ship in the world welcomed her first fare-paying passengers on board. Many of them would not survive the ship's maiden voyage. Her name has gone down in history as a byword for disaster. She was the legendary *Titanic*. Her loss led to the introduction of new regulations for ship design and safety at sea.

TYPE: *Olympic*-class ocean liner

LAUNCHED: Harland and Wolff shipyard, Belfast, Northern Ireland, 1911

LENGTH: 882 ft 9 in (269.1 m)

TONNAGE: 52,310 long tons (53,150 metric tons) displacement

CONSTRUCTION: riveted steel

PROPULSION: two reciprocating steam engines and a low-pressure turbine, 46,000 hp (34,300 kW), driving three screw propellers

Titanic was built in the Irish city of Belfast (now in Northern Ireland), which is still immensely proud of the great ship. She was the biggest man-made object that had ever moved across the face of the Earth. A common refrain in Belfast, even today, is "She was alright when she left here!" *Titanic* and her sister ship *Olympic* towered over the streets of workers' houses huddled around the Harland and Wolff shipyard as the two great vessels took shape on adjacent slipways.

Titanic's hull was built from 2,000 steel plates fixed to 300 frames with more than 3 million rivets. The hull was divided into 16 watertight compartments, sealed by electrically operated watertight doors. They could be closed by a flick of a switch on the bridge. Alternatively, they could be closed individually. If the control system failed, water flooding into a compartment would automatically trigger the door closure. And *Titanic* could stay afloat with two compartments full of water. It seemed inconceivable that any foreseeable accident could sink her.

Titanic's public spaces and passenger facilities were unsurpassed. She had a swimming pool, Turkish baths, squash courts and a gym. There were spacious dining rooms, lounges and a reading room. Her first-class and second-class cabins were decorated and furnished to the highest standards. The aim was to give passengers the same experience as a stay in a well-appointed hotel. Even in third class, or steerage, conditions were better than in the third class of most liners at that time.

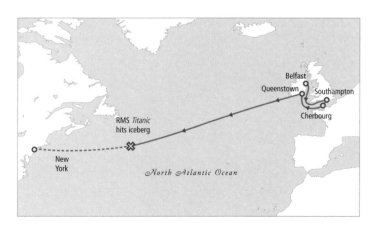

RIGHT: *The route of Titanic's maiden voyage from Southampton to New York, which included stops in Cherbourg, France, and Queenstown (now Cobh), Ireland.*

ABOVE: Titanic *was built to the highest level of opulence, replicating the look and feel of Europe's finest hotels.*

Titanic's third-class passengers were accommodated in cabins instead of large, open dormitories. The ship famously had a grand staircase that extended down through seven decks and was topped with a glass dome.

She was equipped with 20 lifeboats with a total capacity of 1,178 people, but *Titanic* had more than 3,000 passengers and crew on board. The regulations in 1912, when *Titanic* made her maiden voyage, required a ship of her size to have only 16 lifeboats, so she actually exceeded the minimum safety standards. The small lifeboat capacity was thought to be adequate because the lifeboats would be used to ferry passengers between *Titanic* and rescue ships in the event of an emergency. They were not expected to accommodate all of the passengers and crew at the same time.

A MATTER OF CHANCE

When ships as big as the *Titanic* maneuver in a river or port, they can cause suction effects in the water powerful enough to pull nearby vessels toward them. On September 20, 1911, the Royal Navy warship HMS *Hawke* collided with *Titanic*'s sister ship, *Olympic*, in the Solent (a strait on the south coast of England), possibly because she had been sucked toward the giant liner as she passed. *Olympic*'s repairs delayed the *Titanic*'s completion. When *Titanic* finally left for her maiden voyage, she almost collided with the SS *City of New York*. The passing *Titanic* tore the *New York* from her moorings and sucked her toward the *Titanic*, which had to take avoiding action. It is interesting to speculate that if the *Olympic* had not suffered her collision with the *Hawke* or if *Titanic* had collided with the *New York*, *Titanic* would have departed on a different date and probably would not have hit the iceberg that sank her.

ABOVE: *In one of the last photographs of* Titanic *ever taken, the great ship is seen leaving Southampton on her maiden (and last) voyage. Many of the passengers and crew would be dead barely 5 days later.*

The Maiden Voyage

Unlike most other British ocean liners, which were based in, and departed from, Liverpool, *Titanic* was based in Southampton on England's south coast. She was so big that a new dock had to be built for her. The new location was chosen for hard-headed commercial reasons: Southampton had better transport connections with London and was closer to the Continent, where *Titanic* could pick up extra passengers before setting out across the ocean.

As *Titanic* got under way on the morning of April 10, 1912, more than 1,300 millionaires, playboys, businessmen, emigrants, and their families looked forward to the great adventure of a transatlantic crossing. They included Thomas Andrews, Harland and Wolff's managing director and J. Bruce Ismay, chairman of *Titanic's* owners, the White Star Line.

On April 14, *Titanic* started receiving ice warnings from other ships. Captain Smith changed course to a more southerly route to avoid icebergs. Lookouts were posted, but the ship continued at full speed. It was normal practice not to slow down until ice was actually spotted. The sea was exceptionally calm, making it difficult for the lookouts to discern the horizon.

At 11.40 p.m., lookout Fred Fleet in the crow's nest shouted into his telephone, "Iceberg right ahead." Officers on the bridge immediately

RIGHT: Titanic *was commanded by Captain Edward J. Smith. He stayed aboard and perished when the ship sank, having previously said that he would go down with his ship if the worst ever happened.*

I CANNOT CONCEIVE OF ANY VITAL DISASTER HAPPENING TO THIS VESSEL. MODERN SHIPBUILDING HAS GONE BEYOND THAT.

Captain Edward Smith,
***Titanic's* commander**

reversed the engines and turned the wheel hard to starboard, which swung the bow to port. Their quick action prevented a head-on collision, but the ship struck the iceberg a glancing blow. Most of the passengers felt no impact at all. Initially, it looked as if disaster had been averted. However, down in the bowels of the ship, workers in boiler room 6 heard a thunderous roar and were struck by a torrent of ice-cold seawater bursting in through a gash in the hull plates. Officers dashed below to inspect the damage. They were horrified by what they found. Five watertight compartments had been breached and were filling with water. It was immediately clear to them that the ship was doomed. *Titanic* was going to sink.

THE ICEBERG

The iceberg that struck the *Titanic* started as snowfall on Greenland 15,000 years ago. The weight of more snowfall compressed the underlying layers and changed them to ice. The ice slowly slid downhill to meet the sea in the Ilulissat icefjord. In the early 1900s, this fjord gave birth to one or two giant icebergs every year. In 1909, the *Titanic* iceberg was one of them. Weighing up to a billion tons, it would have taken a year to reach the end of the fjord. By then, its weight would have halved. By 1911, it had entered the west Greenland ocean current that carried it down the northeast coast of Canada and into the Atlantic. After the fateful collision, it drifted south and melted.

RIGHT: *Could this be the iceberg that sank the* Titanic? *It was photographed on April 15, 1912, close to* Titanic's *last position.*

BALLARD'S SECRET MISSION

When Dr. Robert Ballard decided to try to find the *Titanic*, he needed help from the U.S. Navy, which was very interested in Ballard's robot submersible technology. The Navy made a deal with him: If he would use his technology to investigate the wrecks of two American submarines, the USS *Thresher* and *Scorpion*, and if any of the time allocated to the mission remained, he could use the Navy's resources for his own purposes. The *Thresher* and *Scorpion* had sunk in the 1960s and the Navy wanted to know what had happened to their nuclear reactors. Once Ballard had found the wrecks of the submarines and surveyed them, he was left with only 12 days to find the *Titanic*. He had noticed that debris from the submarines was spread over a wide area of seabed and surmised that *Titanic* would be surrounded by an even bigger debris field, so he searched for this, not for the ship itself. And he found it on day 10, September 1, 1985. It led him straight to the ship.

BELOW: *The* Titanic's *bow looms out of the darkness in the deep ocean, with "rusticles" of decaying iron hanging from her.*

Distress messages were sent out, but the nearest ship, *Californian*, did not receive them because her wireless room had closed for the night. Rockets fired from *Titanic* were seen by the *Californian*, but there was confusion about what they meant and so the ship failed to respond. The closest vessel that did respond was the *Carpathia*, 58 miles (93 km) away. She immediately headed for *Titanic*'s position. Meanwhile, officers on the *Titanic* were having trouble persuading passengers, who could not believe that the ship really was sinking, to get into the lifeboats. Some of the boats departed half-full. Eventually, the ship's fate became obvious to everyone. As she continued to sink by the bow, her stern reared up out of the water. Then the hull split in two and slid below the waves. Captain Smith, an old-school sea captain, went down with his ship.

The *Carpathia* arrived just after dawn and picked up 705 survivors from the boats. The *Californian* finally learned what had happened and arrived on the scene later, just as another ship, the *Mount Temple*, also arrived; but it was too late for more than 1,500 passengers and crew who had already died in the icy water. They collected several hundred bodies, which were taken to Halifax, Nova Scotia. Dealing with so many bodies was a mammoth logistical task for the authorities. Relatives traveled to Halifax from all over North America to try to identify them. Then arrangements had to be made to transport bodies to their hometowns for burial. About a third of the dead remained unidentified and were buried in Halifax.

ABOVE: *News of the* Titanic's *sinking was greeted by an outpouring of public grief. The fact that the ship broke in two before going down was not well known in the immediate aftermath.*

The Aftermath

Inquiries into the disaster were held on both sides of the Atlantic. The result was the International Convention for the Safety of Life at Sea. New regulations covering the numbers of lifeboats carried by ships, the use of rockets as distress signals and 24-hour manning of radio equipment were introduced. In addition, the International Ice Patrol was created to monitor icebergs that could present a danger to shipping.

Searches for the wreck of the *Titanic* in the early 1980s failed to find anything. Then in 1985 an expedition led by Dr. Robert Ballard finally discovered the wreck lying in pieces on the seabed at a depth of 12,460 feet (3,800 m).

The condition and the positions of the two main parts of the ship, 1,970 feet (600 m) apart, surrounded by hundreds of thousands of pieces of debris, indicated that the *Titanic* had split in two at or near the surface and the two parts of the ship then made their way separately to the seabed.

Titanic's sister ships had mixed fortunes. The *Britannic* was employed as a hospital ship during World War I. As she steamed through the Kea Channel in the Aegean Sea in 1916, an explosion holed her below the waterline. She sank in only 55 minutes. The explosion is thought to have been caused by a mine laid by a U-boat. *Britannic* was the biggest ship to be sunk during World War I. By contrast, *Olympic* enjoyed a long career as a passenger liner on the Atlantic. By the time she was taken out of service in 1935, she had made 257 round trips in 24 years, carrying 430,000 passengers a total of 1.8 million miles (2.9 million km).

U-21

By the outbreak of war in Europe in 1914, several navies were operating their first submarines for coastal defense and protecting harbors. Germany was more daring, developing long-range, oceangoing submarines: *unterseebooten*, or U-boats. The vessel that, more than any other, established the U-boat as a deadly fighting machine was *U-21*. It was the first U-boat to sink a ship in World War I, the first submarine ever to sink a ship with a self-propelled torpedo and the first submarine to sink a ship and survive.

TYPE: Type *U-19* submarine

LAUNCHED: Danzig (now Gdansk), 1913

LENGTH: 210 ft 6 in (64.15 m)

TONNAGE: 640 long tons (650 metric tons) displacement surfaced, 824 long tons (837 metric tons) submerged

CONSTRUCTION: riveted steel

PROPULSION: two MAN 8-cylinder 2-stroke diesel engines and two AEG double motor dynamos

*O*ne obstacle stood in the way of Germany's ambition to become a global power in the early 1900s — Britain. At that time, Britain had the world's most powerful navy and intended it to remain so. Germany's battleship-building program represented a threat to Britain. Germany could not win a direct confrontation with the Royal Navy, but calculated that if the German fleet could achieve a certain size and strength, Britain would avoid confrontation and agree to coexist with a more powerful Germany. However, if Britain did attempt to control and limit German naval excursions by blockading the North Sea, Germany had to be able to conduct limited naval battles against the British. Submarines would give Germany an advantage over the more powerful Royal Navy.

The Secretary of State for the German navy, Admiral Alfred von Tirpitz, refused to commit substantial government funds to submarines unless he saw proof that they could operate beyond coastal waters. As a result, Germany quickly developed large, seagoing submarines capable of waging war on deepwater shipping.

RIGHT: *U-21 is on the right, farthest from the jetty, of the front row of this group of U-boats docked at Kiel, Schleswig-Holstein, on February 17, 1914.*

THE SUBMARINE COMES OF AGE

On July 16, 1914, German submarine *U-9* did something totally new: her crew successfully reloaded her torpedo tubes while she was submerged. This was vital to developing the submarine into a practical and effective weapon of war. The implications of this became clear in September 1914 when *U-9* discovered three British cruisers (HMS *Aboukir*, *Hogue* and *Cressy*) guarding the eastern end of the English Channel and sank them all within an hour. Almost 1,500 British seamen died. *U-9* could do this without surfacing by reloading her torpedo tubes

underwater. Navies that had dismissed the submarine as a novelty, not a serious weapon system for combat on the high seas, were shocked by this event and immediately realized how dangerous submarines really were.

RIGHT: *U-Boat* U-9 *returns to harbor, saluted and cheered by fellow sailors after its devastating attack on three British warships.*

U-21 was one of four Type 19 submarines that were built between 1910 and 1913. They were the first German submarines to be powered by diesel engines. They had a range of 7,600 miles (12,230 km), enabling them to cross the Atlantic and return to Germany without refueling, in theory at least. *U-21* had the distinction of being the first submarine to sink a ship with a self-propelled torpedo. She carried four torpedoes that could be fired through two bow tubes or two stern tubes. When the submarine was on the surface, the crew could also open fire with her 3.5-inch (88-mm) deck gun.

U-21 went hunting for British ships on a series of patrols in the Dover Strait and in the north of Scotland during August 1914, but failed to find any suitable prey. On September 5, 1914, she was on the surface near the Isle of May off the east coast of Scotland when her crew spotted smoke in the distance. *U-21* dived and prepared to attack, but the ship, the scout cruiser HMS *Pathfinder*, steamed away. The submarine's crew thought they had lost another target, but then *Pathfinder* turned around and steamed back toward them. *U-21*'s captain, Otto Hersing, fired a torpedo, which struck the *Pathfinder* and detonated one of her magazines. The ship exploded and quickly sank. Some survivors were picked up, but 261 seamen died.

ABOVE: *This painting by Willy Stöwer shows* U-21 *preparing to attack a large passenger ship, the* Linda Blanche, *in the Irish Sea on January 30, 1915. The image exaggerates the size of the* Linda Blanche, *which was actually a tiny coastal steamer.*

U-21 carried on sinking ships up and down the North Sea and English Channel until she was posted to the Mediterranean in April 1915. There, she supported Germany's ally, Turkey. She sank the British warships *Triumph* and *Majestic* off the Gallipoli Peninsula, resulting in the removal of all Allied capital ships to safer anchorages farther away. The Kaiser awarded *U-21*'s crew the Iron Cross for their efforts.

At that time, Italy was at war with Austria-Hungary, but not with Germany. To enable *U-21* to attack Italian shipping, she was commissioned by the Austro-Hungarian Navy and served as that navy's *U-36* until Italy declared war on Germany in August 1916. When she spotted what appeared to be a merchant ship off the Sicilian coast and attacked it, it turned out to be an armed ship in disguise — a Q-ship. It opened fire, forcing *U-21* to dive to escape destruction.

U-21 was ordered back to the North Sea in 1917 when Germany resumed unrestricted submarine warfare after a 2-year suspension following the sinking of the liner *Lusitania* (see page 134). In April

RIGHT: *The U-boat threat was used to sell Liberty Bonds using posters like this. The bonds were sold to the public to raise cash for the war effort. Buying them was seen as a patriotic duty.*

alone, 500,000 tons of British merchant shipping was sunk by U-boats including *U-21*. Merchant ships would have been safer traveling together in convoys protected by warships, but Britain did not have enough warships for a strong convoy system. When America declared war on Germany in April 1917, American ships became available and so convoys were immediately introduced. Even then, Hersing had an answer. On one occasion he maneuvered *U-21* into the middle of a convoy, fired two torpedoes and then successfully evaded the depth charges dropped by the escorting destroyers. In general, however, the convoy system and the rapid introduction of hundreds of antisubmarine ships overcame the U-boat threat.

U-21 ended her career as a training vessel, schooling the next generation of German submariners. At the end of the war, she was due to be handed over to Britain, but she sank while being towed across the North Sea. She had sunk 40 British, Italian, Dutch, Russian, French, Norwegian, Portuguese and Swedish ships, making her one of the most successful U-boats of the war in terms of tonnage sent to the bottom. By the end of the war, Germany had built 375 U-boats of 33 different classes and shown the world what a powerful weapon the submarine could be.

Q-SHIPS

In spring 1916, *U-9* attacked a ship that appeared to be an unarmed merchant vessel. However, it was actually a Q-ship, an armed ship disguised in order to lure submarines into attacking it. The Q-ship looked like a small unarmed freighter or fishing boat. When a U-boat surfaced nearby and warned the ship (in accordance with international treaties) that it was about to be sunk, the crew would take to the boats and leave the vessel. The U-boat would then approach the apparently abandoned ship to open fire with its deck gun. As it closed in, a hidden crew on the ship would open fire on the submarine. Initially, Q-ships were very successful. Britain and France deployed hundreds of them, but they became less effective as time went on. U-boats dealt with them by simply standing off at a safe distance and sinking them with torpedoes. In addition, once unrestricted submarine warfare resumed, U-boats stopped issuing warnings before sinking ships.

BELOW: *This 1918 newspaper image shows a U-boat about to attack a yacht as the crew rows away. The story reveals the yacht to be a Q-ship, so it is the U-boat that is in danger.*

SS NORMANDIE

A change in U.S. law in 1921 led to the construction of a new generation of transatlantic passenger liners. The most beautiful of them all was the SS *Normandie*. She had a revolutionary hull shape and advanced turbo-electric engines. She set new standards for ocean transport. Every liner that followed was compared to the *Normandie*.

FIFTY SHIPS THAT CHANGED THE COURSE OF HISTORY

TYPE: ocean liner

LAUNCHED: Saint-Nazaire, France, 1932

LENGTH: 1,029 ft (313.6 m)

TONNAGE: 70,175 long tons (71,300 metric tons) displacement

CONSTRUCTION: riveted steel

PROPULSION: four turboelectric engines, 160,000 hp (119,300 kW), driving four screw propellers

Great ocean liners like the *Mauretania* and *Olympic* were known for the opulence and luxury of their first-class accommodation, but they relied on carrying large numbers of poor third-class passengers emigrating from Europe to the United States. In 1921, the U.S. Congress passed the Quota Act, which closed the door to mass immigration. In the year after the act was passed, U.S. immigration fell by half a million. Shipowners responded by building new liners for wealthier travelers.

The new ships were designed to be the most beautiful and desirable transports of delight. The main players were Britain and France. Germany was out of the picture because its old liners had been confiscated by the victors at the end of World War I and there was little money available to build new ones. The most famous postwar German liner was the *Bremen*, built for the Norddeutscher Lloyd shipping line. The first of the new French liners was the *Île de France*, launched in 1927 by the Compagnie Générale Transatlantique, also known as CGT

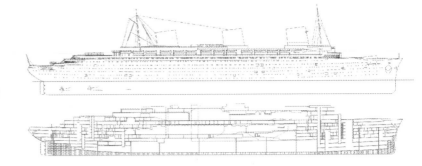

RIGHT: *Normandie's top deck housed her grand saloon and lounges, with the huge first-class dining room below them, and the tourist and third-class facilities to the rear.*

LEFT: *Fifty thousand spectators gathered on the shore to watch the Normandie ease out of Le Havre on her maiden voyage to New York in 1935.*

or simply the French Line. With interiors designed and decorated in the Art Deco style, she was like a floating art installation. She set new standards in design, decor, service and cuisine for the superliners that would follow her. The next French ship, *Normandie*, pushed liner design and style even further. She was breathtakingly beautiful. Her elegant, graceful hull had a gentle upward curve over a concave, clipper-like bow, and her three short funnels leaned back at a jaunty angle. Her bow had an innovative bulbous foot that cut her bow wave and reduced drag.

THE BLUE RIBAND

As soon as one ship crossed the Atlantic, someone else wanted to do it faster. And so it went on. When the great ocean liners were plying their trade across the Atlantic, they vied with each other for the prestige of making the fastest passage. This became known as the Blue Riband, a term borrowed from horse racing. To win the Blue Riband, a ship had to be a passenger liner in regular service traveling westbound at a higher average speed than any other liner. There was no actual prize until 1935 when a British shipowner, Harold Hales (1868–1942), had a trophy made. The Hales Trophy was awarded to the fastest passenger ship crossing in either direction, so it was not precisely the same as the Blue Riband. Thirty-five ships held the Blue Riband over the years, finishing with the SS *United States* in 1952. She was the last, because the record-breakers that followed her were not passenger liners in regular service.

Hard Times

Normandie was designed by a Russian naval architect, Vladimir Yurkevich. Construction began in 1931, but it was a difficult time for shipbuilders following the Wall Street Crash. Several new liners were delayed or canceled. *Normandie's* construction was able to continue only with government assistance. Governments contributed to the cost of these magnificent ships because they were symbols of national pride — and they could be used as troop transports in wartime.

During *Normandie's* sea trials, she clocked up a speed of more than 32 knots (37 mph or 59 km/h) thanks to Yurkevich's sleek hull shape and the use of turboelectric engines. Her turbines did not drive the propellers directly. Instead, they drove generators that powered electric motors, which turned the propellers. This arrangement eliminated the need for the

ABOVE: *These magnificent Art Deco doors and screen separate the smoking room from the grand saloon. The ship's sumptuous interiors featured work by Art Deco masters including René Lalique, Jean Dupas and Emile-Jacques Ruhlmann.*

auxiliary turbines that were normally required to move a liner in reverse. *Normandie*'s electric motors, unlike turbines, could be run in either direction. This saved an enormous amount of weight and machinery space. The liner's Art Deco interiors included vast public rooms. These huge internal spaces were created by splitting the funnel ducts and running them down the sides of the ship so as to leave the central area free. The main dining room was big enough to seat 700 diners.

Normandie took the Blue Riband at her first attempt. Her maiden voyage from Le Havre to New York in May 1935 took 4 days, 3 hours

BUZZED AND STRUCK

Normandie suffered a bizarre accident in 1936. On June 22, Royal Air Force pilot Lieutenant Guy Horsey was engaged in torpedo practice in the Solent, off the south coast of England. Horsey and his fellow pilots were flying their Blackburn Baffin biplanes toward a target and dropping unarmed torpedoes. Meanwhile, *Normandie* entered the Solent to offload mail and passengers before going on to France. Horsey dived toward the target, dropped

his torpedo, turned away and flew down *Normandie*'s port side below funnel height. His engine misfired and he lost control, striking derricks offloading a car and crashing onto *Normandie*'s foredeck. Horsey survived the crash. The captain decided not to delay his departure, and left with the mangled wreckage of the plane still on the deck. The Royal Air Force had to send a team to France to recover the aircraft.

and 14 minutes. She was the first French ship to hold the record, but she only held it for a year before she lost it to her great rival, Cunard's *Queen Mary*. After a refit in 1937, *Normandie* recaptured the record, only to lose it again to the *Queen Mary* the next year.

Despite *Normandie*'s undoubted beauty and speed, she often traveled barely half-full. She gave so much space and emphasis to first-class passengers that she became known as a ship for the rich and famous, not ordinary travelers. The growing numbers of tourists preferred to travel on the *Queen Mary*.

When war broke out in Europe in 1939, *Normandie* was in New York and she was ordered to stay there. In 1941, she was transferred to the U.S. Navy for use as a troop transport ship and renamed USS *Lafayette*. On February 9, 1942, during conversion work for military service, sparks from a welding torch started a fire. The ship's fire protection system had been turned off and disconnected, so the fire spread unchecked. By the time firefighters arrived she was well alight. Fireboats pumped so much water into her to extinguish the fire that she started listing. The list worsened until she capsized. She was righted in 1943 and towed to a dry dock, but the fire damage, deterioration caused by lying on her side in seawater for more than a year and a shortage of skilled workers in wartime meant that she could not be repaired. She stayed in New York until the end of the war, when she was sold for scrap.

Cie G^{le} TRANSATLANTIQUE

French Line

COUPE LONGITUDINALE
DU PAQUEBOT
NORMANDIE
79,280 Tonneaux

LEFT: *In the 1930s, the great liners were widely advertised and promoted by dramatic graphic posters, like this one of the SS* Normandie *by the artist André Wilquin (1899–2000).*

BISMARCK

The German battleship *Bismarck* was one of the most powerful battleships of World War II. She was so dangerous that she was hunted down and destroyed. Despite her speed, her great firepower and her heavy armor, she was defeated by a flimsy "stringbag" biplane from an earlier age. It was an indication that the day of the giant battleship was drawing to a close.

TYPE: *Bismarck*-class battleship

LAUNCHED: Hamburg, 1939

LENGTH: 823 ft (251 m)

TONNAGE: 41,000 long tons (41,700 metric tons) displacement

CONSTRUCTION: welded steel

PROPULSION: three steam turbines, 148,120 hp (110,450 kW), driving three screw propellers

BELOW: *The German battleship* Bismarck *was a formidable fighting machine. From the moment she was launched, she was a serious threat to Allied naval power and merchant shipping.*

In the years between the two world wars, a series of treaties and agreements between the major naval powers attempted to limit warship construction, but these began to break down in the 1930s. When war broke out in 1939, Germany did not anticipate having to fight Britain at sea until the late 1940s, at the end of the forthcoming continental land war. They had a plan, called Plan Z, to achieve naval parity with Britain by then. However, Britain declared war on Germany on September 3, 1939. Germany had only two modern battleships in service, *Scharnhorst* and *Gneisenau*, but two bigger battleships had just been launched and would soon be commissioned: *Bismarck* and *Tirpitz*.

Bismarck was the first to enter service. Her original design called for a main armament of eight 13-inch (33-cm) guns, but by the time she was under construction other countries were building warships armed with bigger 15-inch (38-cm) guns. The German navy wanted to at least match them. *Bismarck* could have mounted even bigger 16-inch (41-cm) guns, but they would have then needed a bigger hull, displacement, draft and magazine, which would have pushed up her construction costs and delayed her completion. As her displacement had already been agreed, it was decided to arm her with eight 15-inch (38-cm) guns in four twin turrets, two forward and two aft. Each of these massive guns could hurl a shell weighing as much as a small car a distance of more than 22 miles (36 km). The weight saved by not increasing the gun size further enabled her to have thicker armor. In case a shell did manage to penetrate the armor, her hull was divided into 22 watertight compartments.

ABOVE: *This painting by Olaf Rahardt, entitled* The Last Battle of the Bismarck, *shows the battleship in action. Shell splashes from the British warships pursuing her are just visible beyond the flash and smoke from her main guns.*

Hunting the *Bismarck*

After a successful raid on merchant shipping in the Atlantic in 1941 by the battleships *Scharnhorst* and *Gneisenau*, it was *Bismarck*'s turn to wreak havoc among the merchant ships supplying Britain. Her sister ship, *Tirpitz*, would normally have joined her, but although she had been launched and commissioned by then she was not yet combat-ready. Instead, *Bismarck* would be accompanied by the heavy cruiser *Prinz Eugen* on what was to be called Operation Rheinübung (Rhine Exercise). If Britain could be starved and blockaded into submission, Germany would achieve complete supremacy in Europe.

Bismarck and *Prinz Eugen*, stationed at Gotenhafen (the former Polish port of Gdynia), were to break out into the Atlantic undetected. *Prinz Eugen* left first, on May 18, and *Bismarck* slipped away the next day. They rendezvoused off Cape Arkona and headed west. When they reached the Kattegat, a shallow sea between Denmark and Sweden, a Swedish warship spotted them. As they continued west and then turned north along the Norwegian coast, they were spotted again by the Norwegian Resistance. The British were now beginning to receive a stream of sightings. On May 21, the two ships sheltered for the day in fjords near Bergen and waited for darkness to fall before moving on. *Prinz Eugen* took the opportunity to refuel. While there, they were discovered and photographed by a Royal Air Force Spitfire.

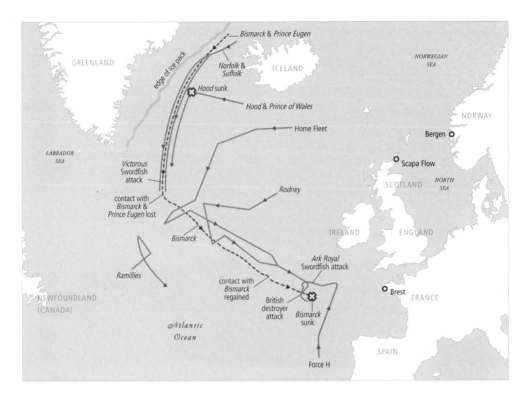

The map shows labels: Bismarck & Prince Eugen, NORWEGIAN SEA, GREENLAND, ICELAND, Norfolk & Suffolk, NORWAY, Hood sunk, Hood & Prince of Wales, Home Fleet, Bergen, Scapa Flow, LABRADOR SEA, Victorous Swordfish attack, Rodney, SCOTLAND, NORTH SEA, contact with Bismarck & Prince Eugen lost, Bismarck, IRELAND, ENGLAND, Ark Royal Swordfish attack, Ramillies, contact with Bismarck regained, British destroyer attack, Bismarck sunk, Brest, FRANCE, NEWFOUNDLAND (CANADA), Atlantic Ocean, SPAIN, Force H, edge of ice pack

ABOVE: Bismarck *almost made it back to protected waters after sinking the British warship, HMS Hood, but once* Bismarck *was located, British warships closed in on her. Here, blue lines represent British ships or fleets and the red dashed line shows the* Bismarck*'s route.*

> WE WILL FIRE UNTIL THE BARRELS GLOW RED-HOT AND UNTIL THE LAST SHELL HAS LEFT ITS BARREL.
> **Admiral Günther Lütjens to *Bismarck*'s crew before their final battle**

Shortly after they left under cover of darkness, more British aircraft arrived and searched for them in vain. By then, the ships were in the open sea and heading due north at speed.

A British battle group left Scapa Flow in Scotland's Orkney Islands on the evening of May 22 to search the routes they thought the German battleships might take from their last known positions. Meanwhile, *Bismarck* and *Prinz Eugen* had turned west. Their commanders still thought their departure from the Baltic had not been noticed; they didn't know that British warships were actively searching for them. It was vital that they be found, because 11 convoys were about to set out across the Atlantic bringing essential supplies and thousands of troops to Britain. HMS *Suffolk* finally spotted the German ships in thick fog on the evening of May 23. The main British battle group, including the new battleship *Prince of Wales* and the aging battlecruiser *Hood*, was about 300 miles (480 km) away. *Bismarck* caught sight of another British ship, HMS *Norfolk*, and opened fire, but *Norfolk* disappeared into the fog. The British ships shadowed *Bismarck* and *Prince Eugen* at a safe distance while they waited for the rest of the British fleet to arrive.

On the morning of May 24, *Bismarck*'s hydrophone (underwater microphone) operators and lookouts detected two ships approaching. One of them, HMS *Hood*, opened fire on *Prinz Eugen* at a range of 25,000 yards (22,860 m). *Hood* was such a powerful and important ship that she was known as "the Mighty *Hood*" and "the Pride of the Royal Navy." Seconds later, the second British ship, *Prince of Wales*, opened fire. Both German ships returned fire, but there were no hits during these opening salvos. The *Prince of Wales* scored the first hit against *Bismarck*. The gunners on both sides had got their range. The *Hood* and *Bismarck* suffered further hits. Then a shell fired by *Bismarck* penetrated *Hood*'s armor and hit one of her magazines. The ship exploded in a shattering blast that broke her in two. She sank within 3 minutes with all but three of her 1,418-man crew. News of her loss and the manner of her sinking caused profound shock in Britain.

The battle continued. The *Prince of Wales*, now drawing fire from both German ships, was hit repeatedly and broke off the attack. *Bismarck* was too badly damaged to continue with Operation Rheinübung and set a course for a French port for repairs. Every British ship available was ordered to find *Bismarck* and destroy her. Ships were brought from as far away as Gibraltar to join the search. HMS *Victorious* launched her aircraft. When *Bismarck* spotted them, she started zigzagging. One of the torpedoes dropped by the planes hit *Bismarck* amidships and forced her to cut her speed. The badly damaged *Prince of Wales* was able to catch up with *Bismarck* and opened fire on her, but the British ships soon lost contact with her again.

DISCOVERY

After sinking in 1941, *Bismarck* was not seen until 1989, when Dr. Robert Ballard, who had discovered the wreck of the *Titanic*, found her sitting upright on the seabed in 15,719 feet (4,791 m) of water, 600 miles (965 km) from the French coast. When she sank, she had landed on the flank of an extinct volcano, causing a landslide. Her hull had been holed in eight places. Part of the stern had broken away. All four of her main turrets were missing; they had probably fallen out when the ship rolled over on the surface. Apart from that, the hull was in relatively good shape, indicating that the ship may have flooded with water when she sank, equalizing the pressure inside and out. This suggests that she may have been deliberately scuttled to stop her falling into enemy hands, as some of her crew claimed.

The Final Engagement

Bismarck's crew thought they were safe as they approached waters patrolled by U-boats and within range of German aircraft. But a Catalina flying boat found her off the west coast of Ireland on the morning of May 26. She was about 700 miles (1,125 km) from the French port of Brest. The British ships were now so far away (because they had been steaming in the wrong direction) that *Bismarck* could not be caught unless she could be slowed down even more. Fairey Swordfish biplanes from HMS *Ark Royal* attacked her with torpedoes. One of them hit her stern, jamming her rudders to port. *Bismarck* was now crippled and only able to circle. U-boats were in the area, but the sea was too rough for them to assist her. The next morning, May 27, the British ships *Rodney*, *King George V* and *Dorsetshire* sighted *Bismarck* and opened fire. *Bismarck*, a sitting duck, suffered repeated hits by shells and torpedoes. One by one, her turrets were hit and put out of action. Hundreds of her men died in the onslaught as the firing continued unabated. The British were determined to destroy the ship that had sent the *Hood* and her crew to the bottom. Nearly 3,000 shells were fired at the *Bismarck* and up to 400 found their target. The blazing *Bismarck* finally rolled over and sank. *Dorsetshire* started picking up survivors in the water, but left when a submarine's periscope was reportedly spotted. Hundreds of men were still in the water; of the 2,200 men on *Bismarck*, only 115 survived. Operation Rheinübung's commander, Admiral Günther Lütjens, and *Bismarck*'s only commander, Ernst Lindemann, both went down with the ship.

THE FATE OF GERMANY'S CAPITAL SHIPS

Bismarck's sister ship, *Tirpitz*, represented a serious danger to Arctic convoys. On November 12, 1944, the Royal Air Force finally sank her by dropping 12,000-pound (5.4-metric ton) "Tallboy" bombs on her. Other German capital ships were either sunk or blockaded in port. In February 1942 *Scharnhorst*, *Gneisenau* and *Prinz Eugen* broke out of Brest, France, into the English Channel and managed to reach their home ports in Germany. However, *Gneisenau* was bombed while in dry dock and never put to sea again. *Scharnhorst* saw action again in the last naval battle ever fought between British and German battleships, the Battle of the North Cape in December 1943. She attacked a convoy off the north coast of Norway, but was overwhelmed by the convoy escorts and sank. *Prinz Eugen* survived the war and was handed over to the U.S. Navy. She was used in two atomic bomb tests in the Pacific to observe the effects of nuclear blasts on ships. Left to deteriorate, she sank in December 1946 at Kwajalein Atoll in the Marshall Islands.

ABOVE: Prinz Eugen *was anchored near Bikini Atoll in 1946 to test the effects of two nuclear weapons. The tests were codenamed Able and Baker.*

BELOW: Prinz Eugen *was moored 1,200 yards (1,100 m) from the detonation point. She suffered minor damage but was heavily contaminated with radioactive fallout.*

HMS *ILLUSTRIOUS*

The aircraft carrier was, like the submarine, a development that military strategists treated with skepticism. And, like the submarine, it was initially used as a mere support vessel. The potential of the aircraft carrier was finally realized after the Battle of Taranto in 1940. It was the first all-aircraft ship-to-ship naval attack. The ship that demonstrated what a potent weapon the aircraft carrier could be was HMS *Illustrious*.

TYPE: *Illustrious*-class aircraft carrier

LAUNCHED: Barrow-in-Furness, England, 1939

LENGTH: 740 ft (225.6 m)

TONNAGE: 23,000 long tons (23,369 metric tons) displacement

CONSTRUCTION: armored steel hull

PROPULSION: three steam turbines, 111,000 hp (83,000 kW) driving three screw propellers

*I*taly had stationed a powerful fleet of six battleships plus cruisers and destroyers at Taranto on its south coast to protect its supply lines across the Mediterranean Sea to North Africa. Britain had to neutralize this fleet to cut the Italian supply lines and eliminate the threat to Allied military operations in North Africa. The British decided to attack the Italian fleet with torpedoes dropped from naval aircraft. HMS *Eagle*, an old converted superdreadnought battleship dating from 1918, was chosen for the job. However, *Eagle* was found to have a leaking fuel system that needed urgent repair, so HMS *Illustrious* was substituted.

Illustrious, which had been launched just a few months before World War II began, was fitted with early-warning radar. This allowed the ship to detect aircraft up to about 60 miles (100 km) away. Enemy aircraft that came closer than 12 miles (19 km) were engaged by the ship's 16 4.5-inch (110-mm) guns arranged in eight twin-gun turrets, four on each side of the flight deck. If an enemy aircraft got within 6,800 yards (6,200 m), the ship's quick-firing "pom-pom" guns would open up. *Illustrious* was armed with 48 of them arranged as six octuples (eight-barreled units). If an aircraft managed to get through

FIFTY SHIPS THAT CHANGED THE COURSE OF HISTORY

all of that and drop bombs, the ship was protected by an armored flight deck with an armored aircraft hangar below it.

Critics doubted that the attack on Taranto, code-named Operation Judgement, would be successful, and argued for a conventional sea battle instead. Taranto was heavily protected by antiaircraft guns, machine-guns and barrage balloons. There was also some torpedo netting in place. An attack from the air was going to be very difficult.

INVENTING THE AIRCRAFT CARRIER

The first attempts to launch planes from ships took place only 7 years after the Wright brothers' historic first powered flight in 1903. The first successful take-off was made by American pilot Eugene Burton Ely (pictured) in a Curtiss Model D biplane on November 14, 1910. The ship, the USS *Birmingham*, was standing still on a flat calm sea. To be a practical system, planes would have to take off from the pitching and rolling deck of a ship

under way and land back on the same ship again. In May 1912, the first take-off from a ship under way was made from the British warship HMS *Hibernia* by Commander Charles Samson flying a Short Improved S.27 biplane. The first ship capable of launching and recovering aircraft was HMS *Argus* in 1917. Early aircraft carriers were converted warships and liners; the world's first purpose-built aircraft carrier was the Japanese ship *Hosho*, launched in 1921.

BELOW: *The Fairey
Swordfish planes carried
by HMS* Illustrious *were
torpedo bombers.
Nicknamed "stringbags,"
each was armed with a
single torpedo.*

The Battle of Taranto

On November 11, 1940, *Illustrious* and eight other warships steamed to a position off the Greek island of Cephalonia, about 200 miles (310 km) from Taranto. After sunset, six Swordfish biplanes armed with torpedoes, and six more with bombs, took off from *Illustrious*. They were followed by a second wave of nine aircraft, one of which had to turn back because of fuel problems. They divided into smaller groups for the attack. The first group bombed oil tanks on the shore. The next group began the attacks on the ships in the harbor. One by one, the ships were hit by torpedoes or bombs.

By the end of the battle, one battleship (*Conte di Cavour*) had been sunk and another two battleships and two heavy cruisers had been damaged. *Conte di Cavour* was refloated, but she was so badly damaged that repairs were never completed. Of the 20 aircraft that reached the target, two were shot down; the two airmen in one plane died and the two in the other plane were taken prisoner. Italian casualties amounted to 59 dead and 600 wounded. Taranto was a success. It was the first time a fleet had been put out of action by naval air power alone. Military strategists quickly realized that a carrier's aircraft had a much longer reach than a battleship's guns. No more battleships were built after this, and their place as the dominant warship was taken by the aircraft carrier.

BELOW: *Taranto, a naval
stronghold since the 1860s,
was a significant naval base
in both world wars. HMS*
Illustrious *led an attack on
the port in 1940 to neutralize
its threat to Allied shipping.*

ABOVE: *While* Illustrious *was returning to the UK from the USA in December 1941, she collided with HMS* Formidable *during a storm and damaged her bow and flight deck.*

THE WEIGHTY MATTER OF ARMOR

Aircraft carrier design took two different paths in Britain and the United States during World War II. The United States favored an unarmored flight deck, whereas Britain opted for an armored flight deck. Both approaches had their pros and cons. An unarmored flight deck was quicker and easier to repair, and the weight saved enabled the ship to have a bigger hangar for more aircraft. However, if the flight deck was penetrated by a bomb, the damage and casualties could be devastating. An armored flight deck integrated with an armored hangar created an immensely strong structure, but the extra weight of the deck armor reduced the size of the hangar and cut the number of planes that could be carried. As a result of their experience during the war, all American aircraft carriers from the *Midway* class of 1945 onward had armored flight decks.

Illustrious's aircraft continued attacking enemy vessels, shore positions and aircraft in the Mediterranean. In January 1941, she was badly damaged when attacked by a group of German Stuka dive-bombers. She was bombed again while undergoing repairs in Malta. If she had not had an armored flight deck, she would probably have been destroyed. In 1943, she was back in Britain for a refit before leaving for the Mediterranean again to support Allied landings at Salerno. In 1944 she was posted to the Indian Ocean and in 1945 she transferred to the Pacific, where she suffered kamikaze attacks. She was so badly damaged that she had to go to the Norfolk Navy Yard in Portsmouth, Virginia for repairs. The war ended while she was being repaired, after which she became a training ship. Over the next few years she took part in sea trials to launch and recover jet aircraft including de Havilland Sea Vampires, Supermarine Attackers and Gloster Meteors. She was decommissioned in December 1954 and broken up for scrap in 1957. *Illustrious* had been repeatedly modified and improved during her service life, with new guns, better radar and an extended flight deck. She had served as a floating runway for everything from biplanes to jet planes.

SS PATRICK HENRY

When World War II began, massive amounts of materials, supplies and vehicles had to be transported by sea, but there were not enough ships to do it all. In addition, German U-boats were sinking merchant vessels as fast as they could be built. Large numbers of new merchant ships were needed and they had to have a simple, straightforward design so that they could be built quickly and inexpensively. They were called Liberty ships. The *Patrick Henry* was the first of many.

TYPE: EC2 Liberty Ship

LAUNCHED: Baltimore, Maryland, 1941

LENGTH: 441.5 ft (134.6 m)

TONNAGE: 14,245 long tons (14,474 metric tons) displacement

CONSTRUCTION: steel hull

PROPULSION: triple-expansion steam engine, 2,500 hp (1,900 kW), driving a single screw propeller

BELOW: *Liberty ships like the* Patrick Henry *had a very simple design that could be built quickly and cheaply. The hull was divided into five cargo holds.*

After World War I, the United States returned to its previous neutral, isolationist position, determined to avoid any further "foreign entanglements." That would change when Japan attacked the U.S. Pacific Fleet at Pearl Harbor in December 1941. Until then, America made its shipbuilding resources available to Britain to produce the merchant ships it needed. The plan was to build scores of identical ships as quickly as possible. Their design was based on a type of British tramp steamer dating from 1879. Sixty of this type, known as *Ocean*-class ships, were built in American shipyards in Portland, Maine and Richmond, California. Then the design was modified and improved to meet American manufacturing and shipbuilding standards, deal with wartime shortages of some materials and make the ships easier to build quickly and cheaply. The resulting modified design was called the EC2: E for Emergency, C for Cargo and 2 to indicate a medium-sized ship, 400 to 450 feet (120–140 m) long at the waterline. They carried all sorts of cargo, from food and fuel to weapons and ammunition.

The first EC2 was *EC2-S-C1*. The S indicated that she was steam-powered. All the available steam-turbine engines were allocated to the navy, so these early Liberty ships were powered by simple oil-fired reciprocating steam engines. The *EC2-S-C1* was named *Patrick Henry* after the 18th-century American politician. Henry is famous for a speech he made in Virginia in 1775 advocating military

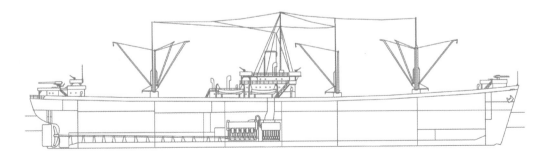

resistance to the British that included the phrase, "Give me liberty, or give me death." President Roosevelt referred to this speech at the *Patrick Henry*'s launch and expressed his hope that these new ships would bring liberty to Europe. As a result, they became known as Liberty ships. They were also known, less flatteringly, as "Ugly Ducklings" and "Sea Cows" because of their utilitarian appearance and slow speed. Although they were not the most attractive ships ever built, they were very successful in service.

Launching the *Patrick Henry*

The *Patrick Henry* was launched at the Bethlehem-Fairfield Shipyard in Baltimore, Maryland, on September 27, 1941, along with another 13 identical Liberty ships. More than 1,000 spectators, including four descendants of Patrick Henry, had gathered in the

ABOVE: *Less than 5 months after her keel was laid, the freshly painted* Patrick Henry *slid down the slipway into the water.*

warm breeze to see the ship named after him take to the water for the first time. She could carry up to 10,685 long tons (10,856 metric tons) of cargo in five holds and on deck: equivalent to 2,840 jeeps, 440 tanks, or 230 million rounds of small-arms ammunition.

To enable them to be built fast, the Liberty ships were assembled from a series of prefabricated sections and modules. The process, similar to assembly-line production in the car industry, was developed by the American industrialist Henry J. Kaiser. Each of the first batch of Liberty ships took about 244 days to build, but within a couple of years this had been cut to just 42 days. In a publicity stunt, a Liberty ship called the *Robert E. Peary* was built from scratch in only 4 days.

Armament was fairly basic and varied from ship to ship, but it often consisted of a 3-inch (75-mm) gun at the bow and/or a 4-inch (100-mm) gun mounted at the stern, with a variety of smaller antiaircraft guns. Crewing Liberty ships was seen as a dangerous job. If they were not sunk by U-boats or aircraft,

I THINK THIS SHIP WILL DO US VERY WELL.

U.S. President Franklin D. Roosevelt, commenting on the design of the Liberty ship

RIGHT: *Nine newly built Liberty ships are readied for service at the California Shipbuilding Corporation in Los Angeles in December 1943.*

BELOW: *The keel of another Liberty ship is laid down. The same process happened more than 2,700 times at 18 American shipyards during World War II.*

or lost in bad weather, they might suffer structural failures, leading to another of the Liberty ship's nicknames: "Kaiser's Coffins."

During the war, the *Patrick Henry* made voyages to the Red Sea, Murmansk in the Russian Arctic, the West Indies, South Africa, West Africa and Italy. She survived the war and was refitted for civilian service, but was badly damaged when she ran aground off the Florida coast in 1946. She was taken to Mobile, Alabama, and laid up with other damaged and retired Liberty ships, but she was never repaired. Finally, she was towed to Baltimore, where in 1958 she was scrapped in the same shipyard where she had been built. The scrap was melted down and turned into steel plate for building new ships.

Victory Ships

In 1944, the Liberty ship design was upgraded. The new ships, known as Victory ships, were bigger and faster so that they could outrun U-boats. They were 455 feet (139 m) long and displaced 15,200 long tons (15,440 metric tons). Instead of the Liberty ship's simple steam engine, most Victory ships were powered by steam turbine engines. A few were diesel-powered. Liberty ships had been prone to hull fractures, sometimes breaking in two in heavy seas.

To remedy this, the Victory ship's hull frames were moved further apart to give the hull more flexibility so that it could bend a little more without cracking. The first Victory ship was the SS *United Victory*, launched in January 1944. A total of 534 were built. A group of 117 Victory ships, called *Haskell*-class attack transport ships, were designed to transport 1,500 troops and land them on coasts using the ship's own landing craft. A handful of them were used as hospital ships. The last of the *Haskell*-class ships was scrapped in 2012.

Forgotten Heroes

The Liberty ships and their courageous crews are unacknowledged heroes of World War II. Their contribution to the war effort was invaluable. Two-thirds of all the cargo that left American ports to support the Allies during the war was carried by Liberty ships. Without them and the essential supplies they ferried across the Atlantic, the U-boat blockade of Britain might have succeeded and changed the course of the war in Europe.

THE FATE OF THE LIBERTY SHIPS

By the end of the war, nearly 3,000 Liberty ships had been built. Most were named after famous patriots, heroes, politicians, scientists and explorers. Two hundred were sunk by enemy action, weather and accidents. Many of them continued in service until the 1960s and 1970s before being scrapped. Only three of the thousands built during the war have survived to the present day. The SS *Jeremiah O'Brien* (pictured) takes passengers for cruises in the San Francisco Bay Area. The SS *John W. Brown* is a museum ship in Baltimore , and also takes passengers on history cruises. The SS *Arthur M. Huddell*, later renamed *Hellas Liberty*, is a stationary museum ship in Piraeus, Greece.

YAMATO

Japan built the largest, heaviest and most powerfully armed battleships of World War II. The biggest was the *Yamato*. However, she proved to be so ineffective in the face of attacks from submarines, surface ships and aircraft that no battleship of comparable size was ever built again. She marked the end of the era of big-battleship navies.

TYPE: *Yamato*-class battleship

LAUNCHED: Kure Naval Dockyard, Japan, 1940

LENGTH: 863 ft (263 m)

TONNAGE: 64,000 long tons (65,027 metric tons) displacement

CONSTRUCTION: welded steel hull

PROPULSION: four steam turbines, 150,000 hp (111,855 kW), driving four screw propellers

The first few months of the Pacific war went well for Japan. Japanese forces took one island after another, but by May 1942 American opposition was beginning to tell. The Japanese advance was stopped at the battles of the Coral Sea and Midway. American forces then pushed the Japanese back across the Pacific. Japan planned a fight-back at sea, assembling a formidable naval force of battleships, aircraft carriers and other ships, including the biggest battleship ever built: the *Yamato*. Her class was planned to have five ships, but only two, the *Yamato* and *Musashi*, were completed as battleships. A third ship, the *Shinano*, was converted into an aircraft carrier.

The *Yamato*, conceived in the early 1930s, was designed to be bigger than any existing Pacific warship and in particular bigger than anything the United States could bring through the Panama Canal into the Pacific Ocean. She was built in secrecy, so her appearance in the Pacific was a surprise. Her armored deck was designed to withstand a direct hit from a 2,200-pound (1,000-kg) armor-piercing bomb dropped from 10,000 feet (3,000 m). Her belt armor was thick enough to cope with an 18-inch (45-cm) shell fired from a distance of 13 miles (21 km). Her main armament consisted of nine 18.1-inch (460-mm) guns in three triple-gun turrets. They were the biggest naval guns ever mounted

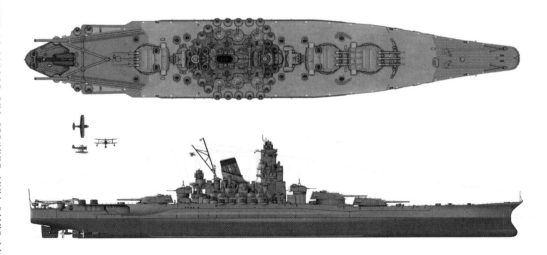

ABOVE: *The* Yamato, *photographed on September 20, 1941, nears completion at the Kure Naval Base on Hiroshima Bay, with the supply ship* Mamiya *in the background.*

on a ship. Each turret weighed a staggering 2,774 long tons (2,516 metric tons), as much as a whole U.S. destroyer. The guns had a range of 27.3 miles (44 km). She was commissioned at the end of 1941, just over a week after the Japanese attack on Pearl Harbor.

The Battle of Leyte Gulf

Despite the *Yamato*'s size and terrifying power, she fired her main guns at another ship in only one engagement, the Battle of Leyte Gulf in October 1944. It was the biggest naval battle of World War II, perhaps the biggest ever.

Nearly all of Japan's carrier aircraft had been destroyed at the Battle of the Philippine Sea in June, 1944. American pilots found it so easy to shoot down the obsolete Japanese aircraft flown by novice pilots that they called it the Great Marianas Turkey Shoot. The *Yamato* was present but took no part in the battle. Then on October 20, U.S. forces invaded the island of Leyte in the Philippines. The battle involved four separate naval engagements between October 23 and 26: the Battle of the Sibuyan Sea, the Battle of Surigao Strait, the Battle of Cape Engaño and the Battle of Samar. Japan sent nearly all of its available warships to Leyte in an attempt to foil the U.S. landings. American submarines spotted the Japanese ships in transit and attacked them, but the *Yamato* escaped unscathed. During her first naval engagement at the Battle of

LEFT: *The Japanese super-battleship* Yamato *was designed to survive ferocious air attack — her central citadel was surrounded by dozens of antiaircraft guns.*

the Sibuyan Sea, she suffered minor damage when she was hit by armor-piercing bombs dropped by aircraft from the U.S. carrier, *Essex*. *Yamato*'s sister ship, *Musashi*, fared less well: she sank after being hit by multiple bombs and torpedoes. On October 24, part of the Japanese force left Leyte and attacked a small group of American ships supporting the landings. This engagement is known as the Battle of Samar, and was the only time the *Yamato* engaged other surface ships. Her guns scored hits on several U.S. ships, but a determined and intense retaliation forced her to leave the area. She returned to the Kure Naval Base in Japan for repairs.

Operation Ten-Go

On April 1, 1945, Allied forces invaded Okinawa. It was a prelude to an invasion of the Japanese home islands and involved the largest amphibious assault of the Pacific war. At the emperor's insistence, the Japanese navy responded by mounting a mission code-named Operation Ten-Go. The *Yamato*, eight destroyers and a cruiser were sent to Okinawa to attack the Allied forces. It was understood by all concerned that this was a last-ditch, one-way suicide mission. The ships were fueled only for the outward voyage because they were not expected to return. If they reached Okinawa, they were to beach themselves and operate as shore batteries, pounding the Allied forces until they were destroyed. The crews were then expected to abandon the ships and fight on land.

Unknown to the Japanese, the Allies had intercepted radio messages about Operation Ten-Go and were prepared for it. Submarines reported on the positions of the Japanese ships. Allied aircraft met the approaching Japanese force on the morning of April 7. Hundreds of Allied aircraft were able to attack the Japanese ships at will because they had no air cover of their own. The attack concentrated on the *Yamato*. It was not long before the bombs and torpedoes found their target. By the end of the attack, the *Yamato* was listing, partly flooded and with several guns out of action. Counterflooding (deliberately flooding the opposite side of the ship) reduced the list.

ABOVE: Yamato's bridge tower carried the ship's air defence radar, surface radar, main armament rangefinder and low-angle fire control director equipment.

RIGHT: *A bomb explodes on the Yamato's deck near the forward gun turret as she flees from the Battle of the Sibuyan Sea, one of the four naval engagements of the Battle of Leyte Gulf, on October 24, 1944.*

There was little respite for the crew before a second attack began in the afternoon. After further hits and flooding, the list to port increased, but no further counterflooding could be used. Soon afterward, a third and final attack began, causing more damage and flooding. Just after 2 p.m., unable to steer, with the list increasing, fires raging and sinking imminent, the crew was ordered to abandon ship. Twenty minutes later, with most of the crew still inside, the *Yamato* capsized. As she rolled over, one of her magazines exploded and she sank, taking about 2,500 of her crewmen down with her. U.S. losses amounted to 12 men. The *Yamato*'s wreck was discovered some 40 years later in 1,120 feet (340 m) of water, 180 miles (290 km) southwest of Kyushu. She was found in two pieces, blown apart by the magazine explosion.

The refusal of Japanese forces to surrender when defeat was inevitable, the willingness of Japan to sacrifice large numbers of its people in kamikaze air raids and futile actions like Operation Ten-Go and the existence of a two-million-strong army ready to defend the home islands all contributed to the American decision to use atomic weapons to end the war. A million casualties were expected if the United States attacked Japan in conventional Normandy-type amphibious landings. On August 6, 4 months after the *Yamato*'s sinking, the first atomic bomb was dropped on Hiroshima, followed by a second attack on Nagasaki 3 days later. Japan surrendered on August 15. World War II was finally over — and so was the era of the giant battleship.

LEFT: *The Yamato steams at its top speed of 27.4 knots (31.5 mph or 50.7 km/h) during full-power sea trials in Sukumo Bay in October, 1941. Following successful trials, the ship was commissioned in December.*

Y A M A T O

RV CALYPSO

In the late 1960s and early 1970s the French underwater explorer, Jacques-Yves Cousteau (1910–97), made a series of documentaries about his work called *The Undersea World of Jacques Cousteau*. One of the stars of this series was the research vessel that carried Cousteau and his fellow explorers. Her exploits engaged a global audience in undersea exploration and environmental concerns, and pioneered a new type of popular science broadcasting.

TYPE: research vessel (former minesweeper)

LAUNCHED: Seattle, Washington, 1942

LENGTH: 139 ft (42 m)

TONNAGE: 360 long tons (366 metric tons) displacement

CONSTRUCTION: Oregon pine hull

PROPULSION: two 8-cylinder General Motors diesel engines, 580 hp (430 kW), driving twin screws

Jacques-Yves Cousteau was a former French naval officer and co-inventor (with Émile Gagnan) of the modern Aqua-Lung that made scuba diving possible. In 1951, he found the ship that would serve as his floating research base for more than 40 years. She was a former Royal Navy minesweeper, HMS *J-826*. She was built in the United States and loaned to Britain under the wartime Lend-Lease scheme. Her hull was made of wood (Oregon pine) to reduce the chance of detonating magnetic mines. She was launched on March 21, 1942, at the Ballard Marine Railway Yard in Seattle, Washington, by Isobel Prentice, the daughter of the shipyard's foreman. Ships like *J-826*, known as British Yard Minesweepers, were not named, but were identified only by their number.

J-826 and her crew of 30 served with the 153rd Minesweeping Flotilla in the Mediterranean. She was based in Malta from 1943. In that year, she took part in Operation Husky, the Allied invasion of Sicily. In 1944 she was redesignated *BYMS-2026* and based in Taranto, Italy. She was decommissioned after the war and laid up in Malta,

where she was sold to a private buyer and converted into a car ferry. Renamed *Calypso G*, she carried up to 11 cars and 400 passengers between Malta and Gozo. In 1950, she was sold again, this time to a millionaire member of the Guinness brewing family, who leased her to Cousteau for just 1 franc a year.

Cousteau moved the vessel, now called *Calypso*, to a shipyard in Antibes in the south of France, where she was converted into an oceanographic research ship. Much of the necessary equipment was donated by private companies and the French navy. The ship's interior was fitted with scientific and photographic laboratories and living quarters for the crew of 28. One novel addition was an underwater observation chamber. Built inside a false bow section, it enabled one person to look out into the underwater world about 10 feet (3 m) below the waterline through glass viewing ports.

In November 1951, *Calypso* embarked on her first expedition. She steamed to the Red Sea so that her crew could study corals there. The expedition discovered several new species of plants and animals. This confirmed Cousteau's belief that the best way for people to understand the sea was to go and explore it themselves. In July 1952, *Calypso* traveled to the islet of Grand-Congloué off the south coast of France, where Cousteau and his team studied the wreck of a cargo ship from the second century BCE. *Calypso*'s divers brought thousands of amphoras and pieces of pottery to the surface for local museums.

ABOVE: *Jacques-Yves Cousteau, photographed here at a conference in 1972 when he was 62, was instantly recognizable to television audiences in the 1970s and 1980s because of his popular marine exploration documentaries.*

The Silent World

To help publicize his work and raise funds, Cousteau wrote a book called *The Silent World*, which was later made into an Oscar-winning documentary film. It introduced Cousteau and *Calypso* to a worldwide audience that was hungry for more news about their adventures and expeditions. Further award-winning documentaries followed. Then in 1968, the series of films he is best remembered for began. They were called *The Undersea World of Jacques Cousteau*. The series ran until 1975. The 36 films followed Cousteau and his divers as they traveled the oceans in *Calypso* and explored the world below the waves. The little ship became famous. Her voyages ranged from the Indian Ocean to Antarctica, and from the Amazon River to the Red Sea.

LEFT: Calyspo *arrives in Montreal on August 30, 1980. Her helicopter sits on the ship's helipad, with the diving saucer on deck behind.*

ABOVE: *In 1972, Calypso visited Antarctica and used its helicopter, a hot-air balloon and divers to explore the continent above and below the ice. It was the first time divers had ventured underneath icebergs and the ice shelf.*

The films Cousteau made pioneered the modern style of presenting science to a general television audience in an entertaining way. Cousteau and *Calypso* also became global icons of the emerging environmental movement. The ship was so famous that singer-songwriter John Denver, one of the most successful recording artists of the 1970s, wrote a song about her. Called *Calypso*, it reached number two in the U.S. Billboard Hot 100 chart. And a make of underwater camera was named Calypso after the ship.

Cousteau equipped *Calypso* with a series of experimental watercraft, including a diving saucer (a minisub that looked like a flying saucer) and underwater scooters. He also had her fitted with a landing pad for a small helicopter. Aft of the helideck, a 3-ton hydraulic crane was installed for launching and recovering the diving saucer and other minisubs. *Calypso* became a floating test bed for all sorts of underwater technology.

The Final Chapter?

On the afternoon of January 8, 1996, *Calypso* was berthed in the Port of Singapore awaiting her next expedition when a barge collided with her and holed her hull at the waterline. She took in water, heeled over and sank in about 16 feet (5 m) of water. The captain and an engineer who were on board at the time were able to escape from the sinking

vessel. She came to rest on her starboard side with parts
of her bridge, masts and helideck visible above the water.

After languishing in the muddy water for 17 days, she was
recovered and brought back to France for repairs. Cousteau
had wanted to replace the ageing vessel for some time and
he felt that this accident would probably hasten the process.
However, he was never to resume his underwater adventures
in *Calypso* or her replacement. He died the following year,
1997, at his home in Paris at the age of 87.

Calypso's owners sold the ship to Cousteau's organization, the
Équipe Cousteau or Cousteau Society. After years of delay due to legal
wrangles and financial problems, work on her restoration began in
2007. The French Maritime and River Heritage Foundation recognized
Calypso's importance by naming her a Boat of Heritage Interest
(Bateau d'Intérêt Patrimonial). The restoration work has been
interrupted by further setbacks and difficulties, and so, as of 2015,
the project has not yet been completed. *Calypso*'s future is not assured.
She could be restored as a floating museum and education center,
but another possibility is that she could be towed out to sea and
scuttled to form an artificial reef for the creatures Cousteau spent
much of his life studying and filming.

BELOW: *A rusting
Calypso stands in a shipyard
in Concarneau, Brittany, in
2007 awaiting restoration
work. Her false bow
section housing the underwater viewing chamber
is clearly visible.*

USS MISSOURI

The *Missouri* was the U.S. Navy's last battleship. While she was being built, along with three other *Iowa*-class battleships, the vulnerability of battleships to attacks by torpedoes and aircraft became clear and so no more were built. The *Missouri* went on to serve in the Korean War and Operation Desert Storm before being decommissioned in 1992 as the world's last battleship. She also hosted one of the most significant and historic events of the 20th century — the ceremony that ended World War II.

TYPE: *Iowa*-class battleship

LAUNCHED: Brooklyn Navy Yard, New York, 1944

LENGTH: 887 ft (270 m)

TONNAGE: 45,000 long tons (45,720 metric tons) displacement

CONSTRUCTION: welded steel

PROPULSION: four cross-compound steam turbines, 212,000 hp (158,000 kW), driving four screw propellers

ABOVE: *The USS* Missouri, *photographed in about 1948, towers over a motor launch full of U.S. Naval Academy midshipmen.* Missouri *and the other* Iowa-*class ships were the fastest battleships ever built.*

The *Missouri*, also known as "the Mighty Mo," was one of four *Iowa*-class battleships built in the 1940s. The other three were the *Iowa*, *Wisconsin* and *New Jersey*. They were designed as fast battleships to escort and protect aircraft carriers, especially the *Essex*-class carriers that were then being built. Their main armament consisted of nine 16-inch (406-mm) guns in three three-gun turrets. Each of the guns could be elevated and fired independently.

The *Missouri* was built just in time to serve in the last few months of World War II. Her first action involved protecting a task force launching air raids against Japan in February 1945. A few days later her massive main guns were providing support for troops landing on Iwo Jima. In March, she was attached to another carrier task force launching air strikes against military installations in Japan and bombarding gun emplacements on the coast of Okinawa in preparation for an invasion. A kamikaze aircraft crashed on her, causing minor

SQUEEZING THROUGH THE CANAL

The *Missouri* was designed to be able to transit the Panama Canal between the Atlantic and Pacific, but she could only just squeeze through. Ships passing through the canal have to negotiate locks that lift the ships 85 feet (26 m) from sea level to the canal level and then lower them again. Initially, the locks were designed to be 94 feet (28.5 m) wide. The U.S. Navy asked for them to be increased to at least 118 feet (36 m) to allow warships to pass through with a comfortable margin on each side. Eventually, a compromise was reached and the locks were built 110 feet (33.53 m) wide. The *Missouri* had a beam of 108 feet 2 inches (33 m).

ABOVE: *The* Missouri *squeezes through the Panama Canal's Miraflores Locks on October 13, 1945, with just a few inches clearance each side.*

BELOW: *Admiral Nimitz, seated on the* Missouri's *deck, signs the Instrument of Surrender that marked the end of World War II.*

damage. The Japanese pilot was given a burial at sea with full military honors by the American seamen. The *Missouri* took part in further attacks on the Japanese coast in May and June before leaving for Leyte in the Philippines. Here, she joined the Third Fleet and then returned to Japan to launch further attacks on industrial targets and protect carriers launching air raids. These combined actions eliminated Japan's ability to operate warships in her own home waters.

After the atomic weapons attacks on Hiroshima and Nagasaki at the beginning of August, the *Missouri* steamed to Tokyo Bay, where military leaders including General Douglas MacArthur, the Supreme Allied Commander, came aboard. A Japanese contingent headed by Foreign Minister Mamoru Shigemitsu joined them and signed the surrender document that marked the end of World War II.

After World War II

The day after the surrender ceremony, the *Missouri* left for Pearl Harbor and then went to New York for an overhaul. In May 1946, she took part in the first major naval maneuvers in the Atlantic since the end of the war. She spent the next year engaged in a series of training and ceremonial tasks before another overhaul in New York. She continued with training and naval exercises until the Korean War began in 1950. By then, the other three *Iowa*-class battleships had been decommissioned, but they were recommissioned to serve in the new war. The *Missouri* was the first of the battleships to reach Korea, where her big guns supported Allied landings. She continued with shore bombardments and carrier protection until March, 1951, when she departed for Norfolk, Virginia. After an overhaul and training, she returned to Korea in October 1952 and took part in shore bombardments until March 1953. After the death of her commanding officer due to a heart attack she returned to the United States and, soon afterward, began a major overhaul that lasted until April 1954. Then on a training cruise to western Europe she was accompanied by the other three *Iowa*-class battleships, the only time the four battleships operated together. By February 1955, she had reached the end of her service career and

ABOVE: *Photographed from the top of one of the Missouri's gun turrets, U.S. Navy planes fly past in formation on September 2, 1945, during the surrender ceremonies in Tokyo Bay.*

THE MOTHBALL FLEET

When the Mighty Mo was decommissioned for the first time in 1955, she was held in the U.S. Navy reserve fleets, also known as the Mothball Fleet. The U.S. Navy does not need to maintain at all times the numbers of ships required to fight a major war. Fewer ships are needed in peacetime, but the navy needs to be able to call upon extra ships quickly in an emergency.

Surplus ships built in wartime are held in several reserve fleets and maintained in a seaworthy state so that they can be reactivated quickly if necessary. In most cases, however, the ships gradually deteriorate and become more and more out of date until they are sold for scrap. One of these reserve fleets off the coast of San Francisco once held more than 300 ships, but by 2017 they will all have been scrapped.

arrived at the Puget Sound Naval Shipyard for decommissioning and storage with the Pacific Reserve Fleet. Moored close to land, she attracted thousands of sightseers who wanted to see the deck where the Japanese surrender was signed.

A Call to Arms

Thirty years after she was decommissioned, the *Missouri* was reactivated and recommissioned as part of a naval expansion program by President Ronald Reagan. By then, much of her World War II equipment was obsolete. It was removed and replaced by updated radar, electronic warfare systems and armaments including Tomahawk cruise missiles. She then embarked on a world tour that involved the first circumnavigation by an American battleship in 80 years.

In the late 1980s, she escorted oil tankers in the Strait of Hormuz during a time of tension with Iran. During the Gulf War in 1991, she fired dozens of cruise missiles at Iraqi targets and bombarded coastal positions in support of Operation Desert Storm to repel the Iraqi invasion of Kuwait. She also helped with mine clearance in the Persian Gulf. She returned to the United States in April 1991. After the collapse of the Soviet Union, the *Missouri* was decommissioned for the last time in 1992 and stored once again at the Puget Sound Reserve Fleet. In 1998 she was taken to Pearl Harbor, Hawaii, and opened to the public a year later as a museum ship.

Navies changed forever after the age of great battleships like the Mighty Mo. The emphasis shifted to the aircraft carrier as a more effective means of projecting power overseas, while also deterring the torpedo and air-attack threats that had neutralized the battleship. The Mighty Mo was the last of her type.

RIGHT: *The* Missouri's *guns blaze as she fires a terrifying 15-gun broadside during maneuvers in the Hawaiian Operations Area in 1987.*

KON-TIKI

The Norwegian explorer Thor Heyerdahl (1914–2002) proposed a theory of human migration in the Pacific that contradicted existing ideas. Heyerdahl's response to the criticism he received was to build a replica of an ancient boat and set off across the ocean to test his theory in practice. His voyage and the vessel he built, a raft called *Kon-Tiki*, became famous, but evidence for his theory had to wait for the invention of genetic analysis.

TYPE: raft

LAUNCHED: Callao, Peru, 1947

LENGTH: 46 ft (14 m)

TONNAGE: unknown

CONSTRUCTION: balsa, pine, mangrove wood, fir and bamboo

PROPULSION: sails

BELOW: *Ocean currents carried* Kon-Tiki *across the Pacific Ocean. Thor Heyerdahl thought ancient peoples from South America might have made the same voyage thousands of years ago.*

In the 1940s, the accepted theory of human migration across the Pacific Ocean held that about 5,000 years ago, people from south Asia migrated east into the Pacific and west into the Indian Ocean. The Norwegian explorer Thor Heyerdahl wondered whether Polynesians could have migrated westward from South America instead of eastward from Asia. No one knew whether such a long voyage was possible with the materials and technology available in South America thousands of years ago. Heyerdahl decided to test his theory by attempting the voyage himself.

Heyerdahl and five companions had a raft built from balsa logs and pine boards tied together with hemp rope. It measured 46 feet (14 m) long by 25 feet (7.5 m) wide. They called it *Kon-Tiki*, the name given by the Incas to their Sun-god. The balsa logs were covered with a bamboo deck and topped by a bamboo cabin with a banana-leaf roof. A sail hung from a 30-foot (9-m) mast held up by an A-shaped frame of mangrove wood. A small topsail could be hoisted above the mainsail and a small mizzensail could be hoisted at the stern. A 19-foot (5.8-m) steering oar of mangrove wood and fir was fixed to the stern. Steering control was also aided by the use of centerboards made of pine planks pushed down between the balsa logs into the water.

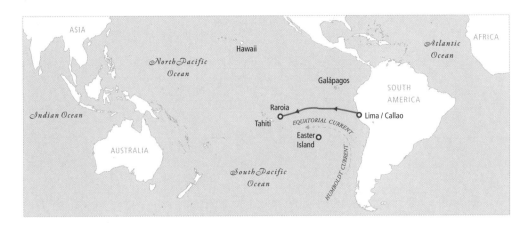

ABOVE: *One of Kon-Tiki's crew fishes for dinner as the raft drifts with the prevailing currents and winds. They caught flying fish, yellowfin tuna, dolphinfish, bonito and shark.*

The supplies the crew took with them included 275 gallons (1,040 liters) of fresh water plus hundreds of coconuts, sweet potatoes and some U.S. Army rations. The crew also caught fish to eat. Several radio transmitters were carried so the crew could report on the raft's progress and alert nearby ships to their presence to avoid collisions.

Just before Heyerdahl set sail, he showed the raft design to some experienced seamen. Their opinion was that the project was doomed to failure. They thought the wooden raft would not stay afloat for the 3 months or so that Heyerdahl had estimated as the minimum duration of the voyage. They thought the broad, blunt-bowed shape of the raft with its tiny sail was impractical, and the ropes holding the raft together would quickly break or rot, causing the raft to fall apart. Other critics were certain that the wood the raft was made from would quickly become waterlogged and sink. Heyerdahl knew that South Americans had used rafts like *Kon-Tiki* to travel great distances along the coast in prehistory, but he didn't know whether *Kon-Tiki* would stand up to a long voyage in the open ocean.

THE *KON-TIKI* EXPEDITION OPENED MY EYES TO WHAT THE OCEAN REALLY IS. IT IS A CONVEYOR AND NOT AN ISOLATOR.
Thor Heyerdahl, from the foreword in the 35th Anniversary Edition of his book, *Kon-Tiki* (Washington Square Press, 1984)

KON-TIKI

Testing the Theory

On April 28, 1947, a Peruvian navy tug towed *Kon-Tiki* from Callao, Peru, into the ocean until it was clear of coastal shipping traffic. Then the Humboldt Current carried her westward into the Pacific Ocean. One hundred and one days later, the raft washed up on a reef at Raroia Atoll in the Tuamotu Archipelago. She had sailed 4,948 miles (7,964 km) at an average speed of 1.5 knots (1.7 mph or 2.8 km/h). She hadn't fallen apart or become waterlogged and sunk.

Although Heyerdahl had successfully demonstrated that South Americans *could* have voyaged to Polynesia in prehistory, most academics still refused to accept that it had actually happened. Years later, scientists were able to use genetic analysis to prove precisely where a people's ancestors had come from. When Polynesians were tested, some of their DNA did indeed originate from South America. Heyerdahl had noticed similarities between the famous carved stone figures called *moai* on Easter Island (Rapa Nui) and pre-Columbian statues from Peru. When DNA samples from Easter Islanders were analyzed, some of them included genes that are also found in indigenous South American people.

Genetic tests showed that most of the Polynesian islanders have Asian origins as was previously believed, but some do appear to be descended from South Americans. Either South Americans voyaged to Polynesia in prehistory, as Heyerdahl had theorized, or perhaps some Polynesians journeyed to South America and then returned home. It is known that some Polynesians did manage to reach South America, because ancient skulls with solely Polynesian DNA have been found in Brazil.

Further Voyages

Half a dozen expeditions successfully repeated *Kon-Tiki's* journey between 1954 and 2011. In the 1960s, a similar raft made an 11,000-mile (18,000-km) voyage from South America to Australia. A raft called *Tangaroa*, whose crew included Thor Heyerdahl's grandson, set off 59 years to the day after *Kon-Tiki* (April 28, 2006) and reached its destination at Raroia Atoll 70 days later, a month faster than *Kon-Tiki*. In 2011, a plastic raft called *Antiki* (because of the age of its crew!) traveled across the Atlantic Ocean from the Canary Islands to the Bahamas, a distance of 3,000 miles (4,800 km). All of these expeditions confirmed Thor Heyerdahl's contention that very long oceanic voyages were possible in the ancient world.

RIGHT: *Thor Heyerdahl and his crew prepare to set out on another expedition, this time aboard a papyrus boat called* Ra.

In a bizarre echo of *Kon-Tiki* in 2010, a yacht called *Plastiki* set sail across the Pacific from San Francisco to Sydney, Australia, to highlight environmental concerns, plastic pollution and the importance of recycling plastic. The 12-ton boat was built from 12,500 reclaimed plastic bottles and other recycled waste plastic.

Heyerdahl himself went on to make further long-distance voyages in replicas of primitive sailing vessels. In 1969, he attempted to sail across the Atlantic Ocean from Safi in Morocco in a boat called *Ra* made of papyrus reeds. During the voyage the reeds became water-logged and sagged until the boat broke up — but it had sailed more than 4,000 miles (6,440 km) and got within about 100 miles (160 km) of landfall in the Caribbean. The next year, Heyerdahl repeated the voyage in another papyrus boat called *Ra II*. This time the voyage was a success and the boat reached Barbados.

Then, in 1977, he built another reed boat, *Tigris*, to prove that the peoples of the three great ancient civilizations in Mesopotamia, Egypt and the Indus Valley could have been in contact by sea. The 60-foot (18-m) boat sailed from Iraq through the Persian Gulf to the Indian Ocean and then to the Indus Valley in Pakistan. From there, it sailed across the Indian Ocean to the Horn of Africa. After 5 months and 4,225 miles (6,800 km) at sea, it finally reached Djibouti. Although it was perfectly seaworthy and could have continued into the Red Sea to Egypt, Heyerdahl burned the boat as a protest against the wars raging in the region.

SS IDEAL X

In the 1950s, a North Carolina businessman lost patience with the traditional way of loading and unloading merchant ships. It was too slow. The solution he came up with revolutionized global trade. The massive containerports that handle the majority of international cargo today, the containerships and the shipping containers themselves can all be traced back to one small ship called *Ideal X*.

TYPE: T-2 oil tanker converted to containership

LAUNCHED: Sausalito, California, 1944; converted Baltimore, 1955

LENGTH: 524 ft (160 m)

TONNAGE: 16,460 gross register tons

CONSTRUCTION: welded steel plate

PROPULSION: steam-turbine electric propulsion

About 90 percent of global trade is carried by ships. Today, the world fleet of merchant ships stands at about 50,000. Every one of these ships has to be loaded at the beginning of its journey and unloaded at the end. The speed and efficiency of the modern cargo terminals that make global shipping possible on this scale are due to a man called Malcom McLean.

In the 1930s, McLean ran a trucking firm in North Carolina. The goods and materials his trucks transported often made part of their journey by sea and McLean was frustrated by how long it took to get the cargo on and off the ships. Special-purpose ships had been developed to handle cargoes such as oil, cars and coal, but almost everything else was transported in general-purpose cargo ships and packed in a variety of sacks, barrels, boxes, bales and crates. It's called "break bulk cargo." When a ship arrived at a port, an army of dockworkers descended on it and unloaded it. Some items had to be handled individually. Others were hoisted on pallets or in nets. Then the ship was loaded with its new cargo in the same way. The whole process took so long that a ship could be in port for weeks. Ships often spent as long in port as at sea.

Finding a Better Way

During Thanksgiving Week in 1937, McLean accompanied a cargo of cotton to New York and watched it being loaded on board a ship bound for Istanbul, Turkey. As the days went by, he grew increasingly frustrated by the delays. Since shipping companies couldn't predict how long the process would take and therefore exactly when a ship would leave one port and arrive at its next port, they had to deliver their cargo to port facilities days or even weeks before it was due to be loaded onto a ship, increasing the chance that some of it might be lost, damaged or stolen. Vast warehouses were needed to store goods at ports, and the process was so labor-intensive that it was also very expensive.

McLean was determined to find a better way. His first idea was to load truck trailers complete with their cargo onto the ships. Then the trailers could be unloaded at their

THERE HAS TO BE A BETTER WAY THAN LOADING CARGO ABOARD SHIP PIECE BY PIECE.
Malcom McLean

destination and driven away. He kept thinking about it and finally came up with an even more radical solution. If haulage companies were to pack all sorts of different goods into identical boxes — shipping containers — the containers could be hoisted on and off ships very quickly.

To test his idea, McLean bought a standard model T2 oil tanker and converted it to carry steel-box shipping containers. Hundreds of T2s were built during World War II. McLean's ship had been launched under the name *Potrero Hills* by the Marinship Corporation of Sausalito, California, on December 30, 1944. McLean installed a new deck called a spar deck above the ship's existing deck. He explained his idea to an engineer called Keith Tantlinger, who designed stackable shipping containers. The containers could be bolted to the spar deck to stop them from moving around at sea. The ship was also renamed *Ideal X*.

At Berth 24 in Port Newark, New Jersey, cranes loaded a single layer of 58 containers onto the new spar deck. Each container took only 7 minutes to load. The whole process was complete in under 8 hours, a fraction of the time conventional loading would have taken. *Ideal X* departed on April 26, 1956, and arrived in Houston, Texas, 6 days later on May 2. The containers were unloaded quickly and driven away by McLean's waiting trucks. The humble World War II tanker had started a revolution.

ABOVE: *Malcom McLean, pictured here aboard one of the vessels he adapted to carry shipping containers, caused a revolution in global freight transport.*

BELOW: *The* Ideal X *was transformed from a tanker to a containership by building a light upper deck, or spar deck, for containers above its existing deck.*

The Next Step

The next year, McLean took the process further. The first containers had to be bolted individually to the ship's deck. McLean modified a new ship, a World War II C-2 class cargo vessel called *Gateway City*, to carry containers stacked on top of each other in racks. The 450-foot (137-m) ship could carry 226 containers, nearly four times as many as *Ideal X*. She was also equipped with her own cranes for lifting the containers.

McLean added more containerships to his fleet, eventually having them designed from scratch as containerships instead of converting existing cargo ships, but the revolution in cargo shipping that he'd started was slow to take off. In the early days, most shipping companies and port authorities thought intermodal freight transport, as this is known, would have limited appeal, so the industry was slow to adopt it. The first container ships didn't cross the Atlantic until 1966.

Eventually, the dramatic cost savings made McLean's shipping containers irresistible. In 1956, he calculated that it cost $5.86 a ton to load a ship in the traditional way compared to only 16 cents when his shipping containers were used. The containers were also more secure, reducing the costs associated with loss, damage and pilferage, which saved even more money. Containerization received an unexpected boost from the Vietnam War. When conventional cargo ships delivered goods and materials intended for American troops, theft at the dockside in

BELOW: *McLean's first containership, the* Ideal X, *has spawned a worldwide fleet of about 5,000 containerships, or about 10 percent of the total merchant fleet.*

RIGHT: *The growth in containerized freight transport has led to the establishment of vast container ports all over the world.*

both American and Vietnamese ports was a severe problem. The use of shipping containers dramatically reduced theft from military cargoes.

Gradually, the inevitable happened. Old, traditional ports and wharves closed, thousands of dockworkers lost their jobs and hundreds of new containerports dedicated to handling shipping containers sprang up around the world. The ships have changed out of all recognition too. Compared to the little *Ideal X*, modern containerships are giants. The biggest can carry more than 19,000 containers. McLean's original containers were 35 feet (just over 10 m) long, because that was the standard truck trailer size in the United States. They were soon standardized to meet international requirements. Modern containers are 20 or 40 feet (6 or 12 m) long.

Having done her job of proving the concept of containerized freight transport, *Ideal X* was sold and renamed *Elemir*. She suffered severe damage in heavy weather on February 8, 1964, and was scrapped in Japan on October 20 of the same year.

WHO WAS FIRST?

Ideal X wasn't the first containership. In 1955, a year before *Ideal X* set sail for the first time, British Yukon Ocean Services launched a freighter called the *Clifford J. Rogers*. Built by Canadian Vickers, she carried her cargoes of general freight and ore between Vancouver and Skagway, Alaska, in 168 containers called "caissons." But it was Malcom McLean and *Ideal X* that spread the revolution in freight transport worldwide.

USS NAUTILUS

Submarines became very effective war machines in the first half of the 20th century, but they were really little more than surface ships that could hide underwater for a while. Then a development in technology in the 1950s transformed the submarine into an altogether more deadly weapon. *Nautilus* was the first of these new submarines.

TYPE: nuclear submarine

LAUNCHED: Groton, Connecticut, 1954

LENGTH: 320 ft (98 m)

TONNAGE: 3,533 long tons (3,590 metric tons) displacement on surface, 4,092 long tons (4,160 metric tons) submerged

CONSTRUCTION: welded steel

PROPULSION: nuclear-powered steam turbines, 13,400 hp (10 MW), driving two screw propellers

*U*ntil 1954, submarines were tied to the sea's surface by the capacity and duration of their batteries. They could submerge on battery power for a short time — a few hours at most — but then they had to surface and start their diesel engines to recharge the batteries. But in 1951 the United States began building the world's first nuclear-powered submarine, *Nautilus*.

After 18 months of construction work, *Nautilus* was launched into the Thames River at Groton, Connecticut. Eight months later she was commissioned as a serving nuclear-powered vessel. She was under way on nuclear power for the first time on January 17, 1955, and quickly smashed all submarine diving and speed records. On May 10 she traveled 1,381 miles (2,222 km) to Puerto Rico in less than 90 hours, submerged all the way. It was the longest submerged cruise and the

RIGHT: *On January 21, 1954, America's First Lady, Mamie Eisenhower, christened the world's first nuclear-powered submarine, USS Nautilus (SSN-571). Moments later, the vessel slid down the slipway and took to the water for the first time.*

highest sustained 1-hour speed by any submarine up until then. Her top speed submerged was certainly more than 20 knots (23 mph or 37 km/h), but how much more was a military secret.

The part of the design that made the difference was, of course, her nuclear reactor. Built specially for a submarine, it used the natural heat of radioactive decay of its fuel to heat water and generate steam for the turbines. Then the steam was cooled, condensed back to water and reused. As the fuel was not burned in the traditional way, it consumed no oxygen and produced no toxic exhaust fumes; these were the reasons why conventional submarines had to use battery power underwater. As a result, nuclear-powered submarines can stay submerged for longer — months if necessary.

A Steep Learning Curve

It was vital for surface ships to learn how to operate with friendly nuclear submarines and how to fight enemy ones, so *Nautilus* spent more than 2 years working through a series of trials and training exercises with the navy. These included test-firing her torpedoes and undertaking exercises with hunter-killer groups in Narragansett Bay, Rhode Island, and off Bermuda. The World War II techniques that employed aircraft and radar to defeat diesel-electric submarines proved to be entirely ineffective against a submarine that could dive deeper, stay submerged longer and leave a search area faster than any previous submarine. At a stroke, *Nautilus* rendered all existing antisubmarine warfare techniques obsolete.

In February 1957, with 69,000 miles (111,100 km) on the clock, *Nautilus* returned to the yard where she was built to have her nuclear fuel core replaced. It was the first overhaul carried out on any nuclear-powered naval vessel. In May, she traveled to the Pacific to take part in a series of trials and exercises with the Pacific Fleet. Then in August she made her first journey under the polar ice. Submarines normally avoided ice in case they became lost or trapped, because magnetic compasses are very inaccurate close to the North Pole and, under the ice, there is no opportunity to take star sightings. It was very easy for a submarine to stray off course and get lost. Nautilus's greater diving endurance eliminated the risk of running out of power while lost under the ice. Being able to navigate under the polar ice had a specific military advantage: The ability to travel from one side of the polar ice to the other would give U.S. submarines a new route into Soviet waters.

ABOVE: *Looking a little dated now,* Nautilus *was the latest thing in 1950s submarine technology. This is the vessel's fire control system, once top secret, but not any more.*

UNDER WAY ON NUCLEAR POWER.
Signal from USS *Nautilus*, January 17, 1955

RIGHT: *Nautilus's route took her from Hawaii and the Alaskan coast to a point in the Greenland Sea near Spitsbergen, and on to England's Isle of Portland. The 1,830-mile (2,945-km) journey below the polar icecap took 4 days.*

Under the Pole

In 1958, *Nautilus* embarked on her most famous mission, code-named Operation Sunshine. She left Seattle on June 9 and, 10 days later, entered the Chukchi Sea north of the Bering Strait, but she had to turn back when she encountered deep ice in shallow water. She waited at Pearl Harbor until the ice conditions improved. On July 23, she headed north to try again. Her commander, William R. Anderson, planned to make a "straight shot" for the North Pole under the ice, but he found that the pack ice had been blown farther south than usual. He searched for the best route and found it at Point Barrow on the northern coast of Alaska. He found a way into a deep seafloor canyon called the Barrow Sea Valley. As *Nautilus* glided through the uncharted waters beneath the ice cap, video cameras on her hull showed the crew pictures of the ice above them.

On August 3, just 2 days after disappearing under the ice, *Nautilus* became the first submarine to reach the geographic North Pole. The water depth measured there by *Nautilus*, 13,410 feet (4,090 m), was considerably deeper than expected. *Nautilus* also discovered hitherto unknown underwater mountain ranges. She spent a total of 96 hours under the ice before surfacing northeast

CARGO SUBS

When *Nautilus* made her historic voyage under the polar ice, commentators at the time thought it presaged a new age of nuclear-powered cargo submarines plying their trade on these new underwater seaways. The distance from London, England, to Tokyo, Japan, via the Panama Canal is about 12,890 miles (20,740 km); the same journey under the North Pole is only 7,250 miles (11,670 km) — not much more than half the distance. Cargo submarines would also be immune from bad weather on the surface. However, nuclear-powered cargo submarines never came about.

of Greenland. She was able to navigate successfully underwater for long periods by using an inertial navigation system adapted from a unit designed for cruise missiles.

After her historic voyage, *Nautilus* returned to evaluation and training exercises. She also took part in the naval blockade of Cuba during the Missile Crisis of 1962. In 1966, while simulating an attack on a warship, she collided with the aircraft carrier *Essex* and damaged her sail (conning tower). After repairs, she continued to take part in operations designed to develop new techniques for using and detecting nuclear submarines, and testing new sensors, weapons and other equipment. As more modern nuclear submarines came into service, developed with the help of lessons learned from *Nautilus*, she served alongside them.

By December 1978, *Nautilus* had logged a total of 500,000 miles (804,627 km) under nuclear power, but her service career was coming to an end. In May 1979, she arrived at the Mare Island Naval Shipyard in Vallejo, California, for decommissioning. In 1982, she was designated a National Historic Landmark and taken to Groton, Connecticut, to be put on display to the public.

BELOW: *On July 1, 2008, the fast attack submarine, USS* Providence *(SSN-719), broke through the ice above it and surfaced at the North Pole to commemorate the 50th anniversary of Nautilus's historic transit under the pole.*

RAINBOW WARRIOR

During a career that spanned 30 years, the *Rainbow Warrior* traveled the world taking part in protests against a variety of environmentally damaging activities and highlighting dangers to wildlife. Her very public role in Greenpeace campaigns all over the world was brought to a sudden end in the most extraordinary and violent way. It's a story involving spies, bombs and government secrets.

TYPE: fishing trawler

LAUNCHED: Aberdeen, Scotland, 1955

LENGTH: 131 ft (40 m)

TONNAGE: 418 gross tons

CONSTRUCTION: welded steel

PROPULSION: diesel-electric engines driving a single screw propeller, plus (from 1985) 6,670 sq ft (620 m²) of sails

The ship that would become a world-famous icon of the environmental movement started out as a humble fishing boat. She was launched in 1955 as a trawler named *Sir William Hardy*, and was the first British ship to have diesel-electric propulsion. She was owned and operated as a research vessel by a department of the British government, the Ministry of Agriculture, Fisheries and Food. By 1977 she was no longer needed, so the government put her up for sale.

Greenpeace UK, the British arm of the environmental campaigning organization, spotted details of the sale and thought the *Sir William Hardy* could be very useful. It took them 8 months to raise the 10 percent deposit needed to secure the purchase. The balance had to be paid after another 60 days. It looked as though the ship would slip from their grasp because of a shortage of funds, until the Netherlands branch of the World Wildlife Fund offered to help finance a campaign to save the whales. This enabled Greenpeace UK to buy the ship. During a 4-month refit in 1977, the *Sir William Hardy* was transformed into the *Rainbow Warrior*, then relaunched on May 2, 1978. Her crew of 24 was drawn from 10 different nations.

DIESEL-ELECTRIC PROPULSION

The type of diesel-electric propulsion system used by *Rainbow Warrior* can be traced back to an oil barge called *Vandal* built in Russia in 1903. She was built to transport oil on the Volga River and the canals of the Volga–Baltic Waterway. Diesel engines were chosen because the vessel was too small for a powerful enough steam engine. The engines drove a generator, which powered the electric propulsion motors. The main advantage of diesel-electric propulsion, then and now, is that the engines can be placed anywhere in the hull, because there is no mechanical link between them and the propellers. The *Vandal* created a lot of interest and led directly to the construction of the world's first seagoing diesel-electric ship in 1908, a tanker called *Mysl*.

Campaigns on the High Seas

Rainbow Warrior's first appearance on the world stage was a protest against whaling by Iceland. Shortages of money, fuel and equipment caused severe difficulties, but the little ship performed well. By 1981, her diesel-electric engines had reached the end of their life and had to be replaced with a new powerplant. During the refit, engineers discovered that the ship's original hydraulic system had been designed to use whale oil as a lubricant! Further protests and relentless pressure applied to whaling nations resulted in a moratorium on commercial whaling in 1982.

In 1985, France was planning the next of its nuclear weapons tests at Moruroa Atoll in the Pacific. By July of that year, *Rainbow Warrior* had reached New Zealand on its way to the test area to register a protest and, if possible, disrupt the test. To make her more environmentally friendly, she had been fitted with a set of sails that made her the world's biggest ketch-rigged sailing vessel. During previous protests against French nuclear weapons tests, the protesters' boats had been boarded by French commandos, so Greenpeace was expecting some opposition, but no one expected what happened next.

THE WORLD IS SICK AND DYING, THE PEOPLE WILL RISE UP LIKE WARRIORS OF THE RAINBOW . . .

The passage from *Warriors of the Rainbow* by William Willoya and Vinson Brown (Naturegraph, 1962), after which the *Rainbow Warrior* ships were named

Attacking a Warrior

Just before midnight on July 10, the 11 crew members on board *Rainbow Warrior* were suddenly plunged into darkness. Almost immediately there was a loud bang and a sudden roar of water rushing into the ship. The crew thought another boat had run into them. Then there was a second, bigger bang. The crew scrambled out of the ship onto the quayside. Within a few minutes, the flooded ship started heeling over. Then the crew realized that one of their number was missing. There was some confusion about whether or not he had been on the ship. Tragically, Greenpeace photographer Fernando Pereira had indeed been on board that night and had drowned.

It was immediately clear that *Rainbow Warrior* had not been hit by any other vessel. The crew suspected sabotage, because nothing in the ship was capable of causing the two explosions that had sunk her. Divers found that the hull had been torn open below the waterline. They discovered an 8 by 6-foot (2.4 by 1.8-m) hole and recovered parts of limpet mines that had not originated in New Zealand. These were evidence that *Rainbow Warrior* had been the target of a deliberate violent attack, but who had carried it out? France immediately came under suspicion.

Two people claiming to be Swiss tourists were arrested. They turned out to be French secret agents, Major Alain Mafart and Captain Dominique Prieur. After initial denials from France, the French prime minister eventually admitted that the pair were agents of the DGSE (Direction Générale de la Sécurité Extérieure), France's foreign intelligence agency, and that they had been acting on the agency's

NEW WARRIORS

Greenpeace went on to acquire two more *Rainbow Warrior* ships — *Rainbow Warrior II* in 1989 and *Rainbow Warrior III* in 2011 — to continue its environmental campaigns. *Rainbow Warrior II* was a schooner built from the hull of a deep-sea fishing ship built originally in 1957. Greenpeace launched the converted ship 4 years after the first *Rainbow Warrior* was sunk. She was retired in 2011 and replaced the same year by *Rainbow Warrior III*, a purpose-built motor-assisted sailing vessel. Funding for this ship was raised by crowd-funding on the Web, which attracted more than 100,000 donors from all over the world.

BELOW: Rainbow Warrior II, *carrying on the work of the first* Rainbow Warrior, *takes part in a protest in 2008 against Israel's plan to build a coal-fired power station in Ashkelon.*

orders. A tribunal ordered France to pay Greenpeace about $8 million in compensation. The secret agents were put on trial, found guilty and sentenced to 10 years and 7 years respectively. They were transferred to a French military base in French Polynesia to serve their sentences. However, France released them within 2 years, attracting further international criticism. A French spy known to have infiltrated Greenpeace's Auckland offices prior to the attack and a combat frogman who took part in the bombing disappeared after the incident and have never been traced.

Rainbow Warrior was refloated a month later, but she was too badly damaged to be repaired. She was scuttled off New Zealand, between Matauri Bay and the Cavalli Islands, in 1987, to become an artificial reef for marine life and divers to enjoy.

NS *Lenin*

When the first nuclear-powered submarines were launched in the 1950s, they seemed to herald a new nuclear age in marine transport. This seemed to be confirmed when the Soviet Union built the first nuclear-powered civilian ship. It was expected to be the first of many nuclear-powered vessels of all kinds.

TYPE: nuclear-powered icebreaker

LAUNCHED: Leningrad, USSR (now St. Petersburg, Russia), 1957

LENGTH: 440 ft (134 m)

TONNAGE: 15,747 long tons (16,000 metric tons) displacement

CONSTRUCTION: welded steel

PROPULSION: three OK-150 nuclear reactors until 1970, (two OK-900 nuclear reactors after 1970), powering turbogenerators and electric motors driving three screw propellers

ABOVE: *After 30 years in service as the world's first nuclear-powered icebreaker, the Lenin was docked permanently in Murmansk and converted to a museum.*

Russia's northern coastal waters provide the shortest route between the east and west of the country, but every winter the sea freezes and icebreakers are needed to keep the Northern Sea Route open. They use an enormous amount of power to break the thick ice. Icebreakers powered by conventional diesel engines burn several tons of fuel an hour and have to make frequent returns to port for refueling. By contrast, a nuclear-powered icebreaker has a virtually unlimited range. It can stay at sea for months at a time without having to refuel. The USSR took the decision to build a nuclear-powered icebreaker in 1953. The ship, called the *Lenin*, was launched 4 years later. She set out on her maiden voyage in 1959.

Her power was supplied by three OK-150 pressurized-water reactors. Two reactors were used in normal service, with the third being held in reserve. Each reactor contained a core of nearly 8,000 enriched uranium fuel

> OUR NUCLEAR ICEBREAKER *LENIN* WILL NOT ONLY BREAK THROUGH THE ICE OF OCEANS, BUT ALSO THE ICE OF THE COLD WAR.
>
> **Soviet Premier Nikita Khrushchev, during a visit to the United States in 1959**

NUCLEAR SHIPS

The Soviet Union, and later Russia, have built a total of 10 nuclear-powered civilian ships — nine icebreakers and a containership, the *Sevmorput*. More are likely to be built. Most of the icebreakers belong to the *Arktika* class. Each of these 22,600-long-ton (23,000-metric-ton) ships can spend more than 7 months at sea and only has to be refueled every 4 years. Satellites monitor the ice and help the crews to plot the best path through it to ensure that the ships do not head into ice that is too thick for them. As well as keeping frozen sea-lanes open, some of the ships have been used to take tourists to the North Pole. Apart from the Soviet/Russian vessels, there have been only three other nuclear-powered civilian ships: the *Mutsu* (Japan), the *Otto Hahn* (Germany) and the *Savannah* (USA). All three have been retired, leaving Russia as the only country currently operating nuclear-powered civilian ships.

pins loaded into 219 fuel elements. It was a very compact unit measuring just over 5 feet (1.6 m) high and 3 feet (1 m) across.

Water flowed through channels in the core to cool the fuel and carry the heat away. The cooling water entered each reactor at a temperature of 478°F (248°C) and left at 1,520°F (825°C). It was prevented from boiling by being held under great pressure, about 28 atmospheres. This superheated, pressurized water then heated water in a secondary circuit to make steam for the ship's turbines. The turbines powered generators producing electricity for the motors that turned the propellers.

The power provided by the reactors enabled the *Lenin* to cut a path through ice more than 8 feet (up to 2.5 m) thick.

BELOW: *The world's biggest icebreaker is the 50 Let Pobedy (50 Years of Victory), an enormous Russian Arktika-class nuclear-powered vessel.*

ATOMS FOR PEACE

The NS *Savannah* was the world's first nuclear-powered merchant ship. She was launched in 1959, 2 years after the *Lenin*. She was proposed by U.S. president Dwight D. Eisenhower and built to promote the peaceful use of atomic power as part of the president's "Atoms for Peace" program. She was never intended, or required, to be commercially profitable. She was more valuable as an ambassador for peaceful nuclear power. In service, she was considered to be a very attractive ship, but she cost more to operate than a diesel-powered ship, needed a bigger crew, carried less cargo and was more difficult to load and unload. She served as a passenger-cargo vessel and then a cargo-only vessel until she was deactivated in 1971. She had traveled 450,000 miles (720,000 km) and called at 87 home and foreign ports, where she was visited by 1.4 million people.

And instead of burning several tons of diesel oil an hour, it consumed just 1.5 ounces (45 g) of uranium fuel a day.

Internally, the *Lenin* is said to have provided her 243 crew members with relatively comfortable quarters that included a sauna, library, movie theater, smoking room and a music room with a grand piano.

Nuclear Accidents

The *Lenin* suffered one or possibly two nuclear accidents. The secretive nature of the Soviet Union during the Cold War, even in nonmilitary matters, meant that few details were released. During refueling in 1965, some of the fuel elements in one of the three reactors were found to be broken and deformed. An investigation discovered that cooling water had been removed from the reactor core while it still contained fuel. Without coolant, the fuel overheated and the elements buckled and even melted in some cases. Only 94 of the fuel elements

could be removed. The remaining 125 were stuck in the reactor and could not be extracted. The whole core was taken out and disposed of in the sea.

A second accident may have happened in 1967 and also involved a loss of coolant, although this may simply be a different account of the first accident. One of the reactors sprang a leak somewhere in its coolant piping. To find the precise location of the leak, the reactor's biological shield had to be opened. The shield, made of a mixture of concrete and metal shavings, was smashed open with sledgehammers, which damaged the core so badly that it could not be repaired. The whole reactor compartment, including steam generators and pumps, was cut out of the ship.

The *Lenin* was then towed to the Zvezdochka shipyard at Severodvinsk in northern Russia,

ABOVE: *The* Lenin *makes short work of this thin ice. She is powerful enough to break through ice up to about 8 feet (1.6 m) thick. Later icebreakers, like the 50 Let Pobedy, can smash through ice twice as thick.*

where a new reactor unit was installed in 1970. Instead of the three original OK-150 reactors, two improved and more powerful OK-900 reactors were installed. The ship was powered by these reactors until she retired in 1989. She was taken out of service because her hull was found to have worn thin in places after 30 years of icebreaking. She had covered more than 575,000 miles (925,000 km) and cleared a path through the Arctic sea ice for 3,741 ships. Spent fuel was removed from her reactors in 1990 and she was taken to the Atomflot nuclear icebreaker base in Murmansk, where she was moored in the harbor. In 2000, a decision was made to convert her into a museum, which opened to the public in 2009.

Other countries experimented with nuclear-powered civilian ships, but the high cost of these vessels, the potentially devastating conse-quences of accidents involving their nuclear reactors and the cost of decommissioning them safely at the end of their service rendered them uneconomical in most cases. Many more nuclear-powered military vessels have been built, especially aircraft carriers and submarines, but the new age of civilian nuclear ships was not to be.

LEFT: *The* Savannah's *control room resembles the control room of a nuclear power station as many of the systems and controls are the same.*

SS TORREY CANYON

On March 18, 1967, a ship called *Torrey Canyon* made history when she ran aground. The accident made headlines all over the world because *Torrey Canyon* was no ordinary ship. At the time, she was the biggest ship ever to be wrecked — a supertanker full of oil. This was the first of the big oil spills at sea.

TYPE: LR2 Suezmax-class oil tanker

LAUNCHED: Newport News, Virginia, 1959

LENGTH: 974 ft (297 m)

TONNAGE: 61,263 gross register tons

CONSTRUCTION: welded steel

PROPULSION: steam turbine driving a single screw propeller

*O*n February 19, 1967, the *Torrey Canyon* filled up with more than 100,000 long tons (101,000 metric tons) of crude oil in Kuwait and set a course for the Canary Islands. When she arrived on March 14, her commander, Captain Pastrengo Rugiati, was informed that the ship's destination would be Milford Haven in Wales. It was usual practice to confirm the destination as late as possible, so the ship could be sent to the United States, a Mediterranean port or western Europe, depending on the oil market. The captain was instructed to arrive no later than

RIGHT: *Heavy seas whipped up by gale force winds crashed against the grounded* Torrey Canyon *and quickly proved too much for the ship. She broke up and dumped her cargo of oil into the sea.*

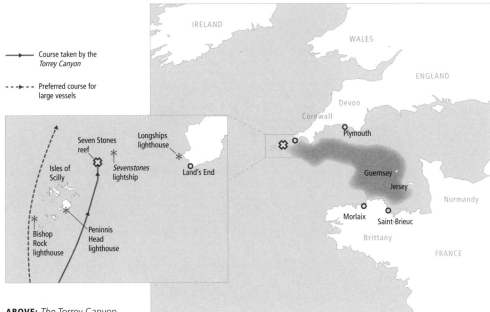

Course taken by the
Torrey Canyon

Preferred course for
large vessels

ABOVE: *The* Torrey Canyon
*ran aground, in spite of local
markers like the* Sevenstones
*lightship, because a
navigational error and
confusion on the ship's bridge
placed her on a collision
course with Pollard Rock
— part of the Seven Stones
reef off the southwest tip
of England.*

11 p.m. on March 18; otherwise it would be another week before
the tide would be high enough again for the ship to enter the port.
It was going to be difficult to get there on time, so the captain was
under pressure to avoid delays.

On March 17, the ship was put on autopilot for the night. In the
morning, Rugiati expected to see the Scilly Isles ahead on his starboard
side, but they were actually to port. Strong ocean currents during the
night had pushed *Torrey Canyon* off course. She was heading for the gap
between the Scilly Isles and the southwest tip of Cornwall, on the
British mainland. Unfortunately, a reef called the Seven Stones lay in
that same gap. The reef, which disappears underwater at high tide,
has wrecked scores of ships.

Shooting the Gap

Rugiati decided to continue through the gap. He intended to steer
down a safe deepwater channel between the islands and the reef.
But the officers then discovered a mistake in their position calculations
— the ship was actually much closer to the reef than they thought.
The helmsman was ordered to make an urgent turn to the north,
but the ship did not respond. The captain thought there must be a
malfunction somewhere. He checked fuses and hydraulics, but there
were no faults. Then he noticed that a control lever on the bridge had
been knocked out of position, disengaging the helm from the rudder.
He reset the lever and ordered a hard turn to port, but it was too late.

The ship hit part of the Seven Stones reef called Pollard's Rock at full speed. The jagged rock ripped the bottom out of the ship.

Oil started pouring into the sea. There was no plan or procedure for dealing with this sort of accident, because it had never happened before on such a large scale. A tug was summoned and preparations were made to pull the ship off the reef. Some of her cargo was pumped out into the sea to lighten her, quickly forming an oil slick 6 miles (10 km) long. The next day, it was more than 20 miles (30 km) long. All the nearby beaches were at risk of pollution, so ships were sent out to spray the oil with detergent to break it up.

Two attempts to refloat the ship failed. Five men were still on board, but an explosion blew two of them into the sea and one died. A few days later, oil reached the English coast. Further attempts to drag the ship off the reef failed. She was stuck fast and quickly started breaking up, dumping more oil into the sea. The decision was made to bomb her and the oil pouring out of her in order to set them on fire. Eight Royal Navy Buccaneer jets dropped 42 bombs, and Royal Air Force jets dropped aviation fuel and napalm, presenting an extraordinary spectacle to onlookers. The ship caught fire, but the oil on the sea would not ignite. There was nothing to stop more of it coming ashore. It polluted beaches in England and France for the next 5 months. Most of the marine life between southern Britain and France was killed and up to 25,000 birds died.

ABOVE: *When a tanker like* the Torrey Canyon *spills oil, seabirds land in it and become covered with it. In an attempt to clean their feathers they swallow the oil. The* Torrey Canyon *spill killed about 25,000 birds.*

ATLANTIC EMPRESS

The biggest oil spill from a ship occurred on July 19, 1979, when the SS *Atlantic Empress* collided with another tanker, the *Aegean Captain*, east of Tobago in the West Indies. Both ships caught fire. Twenty-seven seamen on the two ships died and others were injured. The fire on the *Aegean Captain* was brought under control and her cargo was off-loaded, but the fire on the *Atlantic Empress* could not be controlled. The ship was towed farther out to sea where she sank after a series of explosions, having spilled 282,000 long tons (287,000 metric tons) of oil.

INFAMOUS OIL SPILLS

Two of the most famous oil spills were those of the *Exxon Valdez* and *Amoco Cadiz*. *Amoco Cadiz* was a Very Large Crude Carrier (VLCC) that ran aground on Portsall Rocks off the French coast in 1978. She carried nearly twice as much oil as *Torrey Canyon* and all of it ended up in the sea. *Amoco Cadiz*'s sister ship, MT *Haven*, also spilled her cargo of oil when she caught fire and sank in 1991 off the Italian coast. On March 24, 1989, the *Exxon Valdez* tanker struck Bligh Reef in Prince William Sound, Alaska. She spilled at least 11 million U.S. gallons (41 million liters) of oil into the water, although some estimates give double this amount. The remote location made the incident very difficult to deal with. About 1,300 miles (more than 2,000 km) of coastline and 11,000 square miles (28,000 km) of ocean were polluted.

ABOVE: *When the* Amoco Cadiz *tanker ran aground off the coast of Brittany in 1978, she produced the biggest marine oil spill to that date.*

Learning Lessons

The *Torrey Canyon* disaster was the world's first major oil spill. A review of the rescue and recovery measures found that everything attempted was done too late, was too little or actually made matters worse. Spraying the oil with detergent proved to have been particularly counterproductive. Small oil spills had previously been cleared up this way. However, when the vast slick of *Torrey Canyon*'s oil was sprayed, it made the oil more soluble. This allowed more of it to dissolve in the seawater and thus made it more likely to be taken up by marine organisms. The detergent itself was a problem, too. It might sound like the liquid soap used to wash up dinner plates, but it was actually a cocktail of very toxic chemicals, which proved harmful to marine life.

The *Torrey Canyon* accident provided valuable lessons for those who would have to deal with future disasters. It also led to new international regulations and laws aimed at reducing the risk of further accidents and dealing with them better if they were to happen. And they did happen — since *Torrey Canyon*, there have been at least 10 more supertanker accidents at sea.

USS ENTERPRISE

The USS *Enterprise* (CVN-65) was the eighth U.S. Navy warship to bear the name. She was also the first of a new generation of ships, the nuclear-powered supercarriers. She served with the U.S. Navy for more than 50 years in operations that stretched from the Cuban Missile Crisis in the 1960s to the wars in Iraq and Afghanistan in the early 2000s.

TYPE: *Enterprise*-class aircraft carrier

LAUNCHED: Newport News, Virginia, 1960

LENGTH: 1,123 ft (342 m)

TONNAGE: 93,284 long tons (94,781 metric tons) displacement

CONSTRUCTION: welded steel

PROPULSION: 8 Westinghouse A2W nuclear reactors, 280,000 hp (210 MW), four steam turbines, driving four screw propellers

The design of the *Enterprise* was quite ambitious and experimental. She was powered by eight pressurized-water nuclear reactors built by Westinghouse. Two reactors drove each of the ship's four steam turbines. Two nuclear reactors had never been yoked together like this before, so the engineers designing the propulsion system could not be sure how well it would work. In the end, it was a great success. During a lengthy series of sea trials to test the massive ship's performance, she clocked a top speed of more than 30 knots (35 mph or 55 km/h). She was a floating airport that could be placed anywhere in the world's oceans. Her combined ship and air crew was as big as the population of a small town. In a fully combat-ready state, she carried a ship's crew of 3,200 plus an air wing of 2,480.

One of her first tasks was to act as a tracking station for the first U.S. manned orbital spaceflight by John Glenn in February 1962. A few months later, she was dispatched to her first international emergency, the Cuban Missile Crisis. The Soviet Union had installed nuclear missiles on Cuba, just 90 miles (145 km) from the American coast.

RIGHT: *The USS* Enterprise *(CVN-65) steams through the Indian Ocean in 2003 on her way to the Arabian Gulf to provide air support for Operation Iraqi Freedom.*

RIGHT: *This aerial view of the* Enterprise *shows her island. Originally, it had a domed top carrying radar and other antennae. When the ship underwent a major overhaul between 1979 and 1982, the island was altered to resemble that of* Kitty Hawk*-class carriers.*

The United States blockaded Cuba to stop Soviet ships from delivering more missiles and demanded the removal of the existing missiles. The standoff between the two superpowers lasted for 13 days, during which the world teetered on the edge of nuclear war. Catastrophe was averted when the Soviet Union agreed to remove all the missiles in exchange for President Kennedy's promise not to invade Cuba. Kennedy also secretly agreed to remove U.S. missiles from Turkey.

PROS AND CONS

Nuclear-powered carriers have both advantages and disadvantages compared to nonnuclear carriers. Their main disadvantages are economic. Nuclear-powered carriers are more expensive to build, operate and maintain, and they need nuclear-capable maintenance facilities. But these are outweighed by their advantages. Nuclear-powered carriers have virtually unlimited high-speed endurance, no need for frequent refueling, more electric power for onboard systems, more room for aircraft and aviation fuel and no telltale smoke or exhaust gases to give away their position.

Going to War

After deployments in the Mediterranean, the *Enterprise* was sent to Vietnam. There, she became the first nuclear-powered ship to go to war when she launched air attacks against the Vietcong in November 1965. By the time she left Vietnam in June 1967, her pilots had flown more than 13,400 missions. After an overhaul and crew training, she was sent to Korean waters in January 1968 as a show of force following North Korea's seizure of the U.S. Navy's intelligence-gathering ship, *Pueblo*.

Following repairs for damage caused by a serious fire, she was sent back to Korea in 1969 when North Korea shot down a U.S. Navy reconnaissance aircraft

over the Sea of Japan. Later the same year, she returned to Newport News to have new reactor cores installed, with enough fuel to last 10 years. During the long service life of a nuclear-powered carrier, aviation technology moves on, so that it can find itself hosting aircraft that did not exist when it was designed. In 1973, after another deployment to Vietnam, the *Enterprise* was refitted to host a new naval aircraft, the swing-wing supersonic F-14 Tomcat.

In peacetime, warships are often sent to assist after natural disasters. The vast size and air transport capabilities of carriers are particularly useful, as is the ship's ability to make clean, safe drinking water from seawater. In 1975, the *Enterprise* rendered assistance to Mauritius, which had been devastated by Typhoon Gervaise, before being sent back to Vietnam to deal with a fresh emergency. She took part in Operation Frequent Wind, evacuating personnel from Saigon after an invasion of the South by North Vietnam. During the 1980s, she was deployed to Libya and the Persian Gulf. After another refit and refuel in the early 1990s, she spent the rest of the decade supporting operations in Bosnia and Iraq.

When the United States was attacked on September 11, 2001, the *Enterprise* launched air attacks against Al-Qaeda in Afghanistan. She also provided air support during the invasion of Iraq in 2003–2004 and subsequent operations against Iraq and Afghanistan.

ABOVE: *An F/A-18A Hornet approaches the* Enterprise's *flight deck. The* Enterprise *typically carried about 60 aircraft, but could accommodate up to 90.*

I THINK WE'VE HIT THE JACKPOT.
Admiral George W. Anderson, Chief of Naval Operations, commenting on the Enterprise's **performance during sea trials**

RIGHT: *The USS* Carl Vinson *is one of the current generation of nuclear-powered American carriers, the* Nimitz *class. Lessons learned from building and operating the* Enterprise *contributed to the design of these super-ships.*

A Decommissioning First

Finally in 2012, after a series of port visits that took her around the world, *Enterprise* returned to Newport News, the city where she was built, for deactivation and decommissioning. She was the first nuclear-powered aircraft carrier to be decommissioned and, at 51 years old, she was the oldest active-duty ship in the U.S. Navy. She is scheduled to be scrapped by 2025.

The *Enterprise* was to have been the first of a class of six vessels, but she proved to be so much more expensive than expected that the other ships were never built. She was literally in a class of her own. Instead, she led to a new, improved class of nuclear-powered carrier, the *Nimitz* class. Instead of *Enterprise*'s eight nuclear reactors, *Nimitz*-class ships have just two. Ten *Nimitz*-class carriers were built and all are still active at the time of writing. The first of their replacements, the *Gerald R. Ford* class, has already been launched. Another nine of these nuclear-powered giants are planned to follow. The *Enterprise* name will pass to the third *Ford*-class carrier, *CVN-80*, which is due to be launched in 2023 and commissioned by 2025.

BELOW: *The USS* Rogers *comes alongside the* Enterprise *to render assistance following a serious accident on her flight deck in 1969.*

DSV *ALVIN*

The oceans are the last part of Earth still to be explored. They cover nearly three-quarters of Earth's surface, but more than 95 percent remain uninvestigated. A small submersible craft called *Alvin* revolutionized exploration of the oceans. It has been the preeminent manned deep-diving research craft for more than 50 years.

TYPE: deep-submergence vehicle

LAUNCHED: Bahamas, 1964

LENGTH: 23.1 ft (7 m)

TONNAGE: 20 long tons (20.4 metric tons) displacement

CONSTRUCTION: steel sphere until 1973, then a titanium sphere

PROPULSION: seven battery-powered electric thrusters

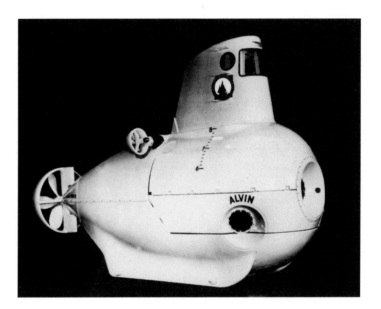

ABOVE RIGHT: *The U.S. Navy released this photograph of* Alvin *in 1964. The little submersible has been upgraded and updated repeatedly since then, enabling it to dive to far greater depths.*

FAR RIGHT: Alvin *was equipped with fine nets for a dive in 1970 to collect plankton for a Harvard University research project.*

Alvin is owned by the U.S. Navy and operated by Woods Hole Oceanographic Institution (WHOI) on the coast of Massachusetts. The institution was formed in 1930, only 54 years after the beginning of modern oceanography with the completion of the *Challenger* expedition. After more than 20 years of research on the ocean's surface, some WHOI scientists started to think about building a new craft to enable researchers to visit the underwater world and see it with their own eyes. The result was *Alvin*, named after WHOI scientist Allyn Vine (1914–94), who had been instrumental in the craft's creation.

Alvin's crew of three (a pilot and two scientists) sit inside a pressure-resistant sphere covered by an outer body that houses the vehicle's air tanks, batteries and a buoyant material called syntactic foam. The crew can move robot arms to pick up samples and place them in a basket fixed to the front of the submersible. Originally, the crew sphere was made of steel, but in 1973 this was replaced by a stronger titanium sphere. It doubled *Alvin*'s depth rating to 12,000 feet (3,660 m). In 1976, tests enabled this rating to be increased again to

THE FIRST DEEP DIVES

The first deep-ocean explorations were carried out in the 1930s by means of a hollow metal ball hanging underwater at the end of a long cable. The ball was called a *bathysphere*. Marine biologist William Beebe and the bathysphere's inventor, Otis Barton, sat inside the ball and looked out through a tiny, thick window. They made more than 30 dives between 1930 and 1934, the deepest reaching a depth of 3,028 feet (923 m). If the cable had snapped, there would have been no way to rescue them. The dives provided the first opportunity for people to witness deep-sea creatures in their natural environment.

RIGHT: *William Beebe looks out through the open hatch of his bathysphere. Its spherical shape resisted the crushing pressure in the deep ocean.*

13,124 feet (4,000 m). In 2011–12, *Alvin* was completely dismantled for a major upgrade that improved its performance and depth rating significantly. A new, bigger and thicker titanium sphere was installed, allowing *Alvin* to dive to 14,760 feet (4,500 m). A second stage of this upgrade will enable the craft to dive to 21,325 feet (6,500 m), bringing 98 percent of the world's seafloor within reach.

The crew can now look out through five viewports instead of three. The old lights and cameras have been replaced with the latest LED lights and high-definition camera equipment. All the thrusters, mechanical arms, batteries and the sample basket were made detachable, so that they can be jettisoned if they should become entangled with something and trap the vehicle underwater. *Alvin* was essentially reborn as a new vehicle.

DSV ALVIN

BELOW: *In 1977, Alvin*
inspected the Galápagos Rift,
an underwater volcanic
hotspot where hot water
gushes out of cracks in the
floor of the Pacific Ocean.

Bombs and Wrecks

Alvin's most famous dives included the search for a lost hydrogen bomb and visits to the wreck of the *Titanic* liner. On January 17, 1966, a U.S. Air Force B-52 bomber collided with a Boeing KC-135 tanker plane during refueling in mid-flight over Spain's Mediterranean coast. Both planes broke up and crashed. The B-52 was carrying four hydrogen bombs. Three landed on the ground near the village of Palomares. The nonnuclear high explosives in the bombs exploded, contaminating the surrounding area with plutonium. The fourth bomb was thought to have landed in the sea. After an extensive search by surface ships and underwater vehicles, *Alvin* finally located the bomb on March 17. An unmanned vehicle called CURV-III (Cable-Controlled Undersea Recovery Vehicle) was sent down to recover it.

In 1977, *Alvin* made some of the earliest dives to hydrothermal vents. These are places where hot, chemical-rich water pours out of the ocean floor, often forming tall chimney-like rock structures known as "black smokers." *Alvin*'s crews were able to observe the marine organisms that cluster around these vents. Then in 1986, *Alvin* made 12 dives to the wreck of the *Titanic*. *Alvin*'s crew photographed the wreck and explored it using a prototype robot vehicle called *Jason*.

Accidents and Angry Fish

Alvin is too small and slow to operate independently like a submarine, so it is described as a submersible. It is carried to its dive sites, launched, and recovered by its own tender or mother ship, the

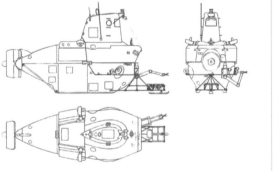

ABOVE: *This three-view drawing of* Alvin *shows the craft's stern thruster, the sail above the crew sphere, the large circular forward viewing port and the sample basket at the front.*

Research Vessel *Atlantis*. A crane on the ship's stern lowers *Alvin* into the sea and brings it back aboard at the end of a dive.

Despite the hostile and dangerous environment in which *Alvin* works, it has suffered very few accidents. In September 1967, a mechanical arm was lost during a rough recovery. The arm was subsequently found, reconditioned and refitted. *Alvin*'s most serious accident occurred in October 1968. While it was being launched, the cable holding it snapped. *Alvin* slid off the ship into the sea and sank in 5,000 feet (1,500 m) of water. The pilot, Ed Bland, who was inside at the time, managed to scramble out before *Alvin* disappeared beneath the waves. The submersible was found and recovered nearly a year later. Remarkably, it had sustained very little damage. The crew's sandwiches inside the sphere were said to be still edible!

During its many dives, *Alvin* has had encounters with all sorts of marine organisms. In 1967, it was attacked by a swordfish while diving in the Bahamas. The fish managed to snag itself in *Alvin's* outer skin and could not get free. It was brought to the surface — and cooked for dinner!

During its 50-year career, *Alvin* has made more than 4,700 dives, enabling 13,000 scientists, engineers and observers to explore the ocean floor. A typical dive lasts about 6 hours — 2 hours to dive, 2 hours' work at depth and another 2 hours to surface again. Alvin can comfortably stay submerged for up to 10 hours, although in an emergency this can be extended to 72 hours. Its dives have revealed hundreds of previously unknown species, from bacteria to giant sea-worms. Its regular upgrades and refits ensure that it will continue to be a leader in undersea exploration for many years to come.

FREE DIVER

The bathysphere led to the development of a new type of diving craft called a *bathyscaphe*. Unlike Beebe's ball, a bathyscaphe is not tethered to a surface ship by cable. Its crew sphere sits underneath a large float full of gasoline. The craft dives by letting seawater flood into tanks to make it heavier. When it is time to come back to the surface, it drops heavy weights to make itself lighter. Small propeller-thrusters give it limited maneuverability. The first bathyscaphe was *FNRS-2* in 1948, but the most famous was *Trieste* (pictured). On January 23, 1960, Jacques Piccard and Don Walsh used *Trieste* to make the first manned descent to the bottom of the Challenger Deep, the deepest known part of Earth's oceans. They reached a depth of 35,814 feet (10,916 m).

GLOMAR EXPLORER

In 1974, a drillship called *Glomar Explorer* headed for a location in the Pacific Ocean to harvest potato-sized chunks of metal, called manganese nodules, that were known to form on the ocean floor. At least, that's the story that was released to the media. In fact, *Glomar Explorer* had been built specially for the U.S. Central Intelligence Agency (CIA) to carry out a daring secret mission to steal military secrets from the Soviet Union.

TYPE: drillship

LAUNCHED: Chester, Pennsylvania, 1972

LENGTH: 619 ft (189 m)

TONNAGE: 50,500 long tons (51,310 metric tons) displacement

CONSTRUCTION: welded steel

PROPULSION: five Nordberg 16-cylinder diesel engines driving generators powering six 2,200 hp (1.6 MW) electric motors turning twin propeller shafts

On March 8, 1968, the Soviet ballistic missile submarine *K-129* sank about 1,560 miles (2,510 km) northwest of Hawaii. The reason for her loss remains unknown. The Soviet navy searched for the submarine but never found her. The U.S. Navy, however, did manage to find her. Their secret underwater acoustic equipment had picked up the sound signature of an "implosion event" at the time when the submarine sank. It was very likely the sound of the submarine being crushed in the deep ocean. This enabled them to plot the approximate position of the event. Then a U.S. submarine, the *Halibut*, scanned the seabed using sonar equipment and cameras until the Russian submarine was found. It was photographed lying on the seabed at a depth of about 16,400 feet (5,000 m).

If the United States could acquire the submarine, it would give them an unparalleled haul of Soviet military secrets. A decision was taken to try to recover the *K-129*. The project was personally approved by President Richard Nixon. The first hurdle to be overcome was that no existing vessel, military or civilian, was capable of salvaging a wreck from such a great depth — so the CIA decided to build one. This raised another problem. If the Soviet Union became aware of an American attempt to snatch one of its submarines, it would likely try to sabotage the mission, so its true nature had to be hidden. The cover story would

RIGHT: *The* K-129 *submarine that the* Glomar Explorer *tried to recover covertly was a Golf II-class Soviet submarine similar to this vessel.*

be that the new ship was merely a run-of-the-mill drillship hunting for valuable manganese nodules. These metal-rich nuggets litter parts of the deep ocean floor. In addition to manganese, they also contain iron, copper, cobalt and nickel. The mission was given the code name Project Azorian.

Building the *Explorer*

K-129 weighed 2,700 long tons (2,743 metric tons) submerged. Salvaging something as heavy as this would require a large ship. Then there was the problem of keeping the true nature of the ship's activities secret. How do you raise a submarine to the surface without a passing spy plane or ship finding out? The solution lay in the innovative design of the ship. Doors in the bottom of the hull would open to allow the submarine to be pulled up into a hidden hold. Then the

ABOVE: *The* Glomar Explorer *returned to her dock at Long Beach, California, in 1974 after her attempt to recover a wrecked Soviet submarine from the seabed.*

hull doors would close and the ship would steam away. The whole James Bond-like operation would take place out of sight underwater.

The *Glomar Explorer* arrived at the recovery site on July 4, 1974. She had a tall derrick similar to an oil-drilling rig. Sixty-foot (18-m) lengths of steel pipe were stored on the deck. A crane lifted each pipe into a vertical position so that it could be screwed into the end of the pipe below it and then lowered into the sea. This continued until the drill pipe was long enough to reach the submarine. A large frame called the capture vehicle was attached to the lower end of the drill pipe. The crew nicknamed it *Clementine*. This had been built in secret, permanently under cover so that spy planes and satellites could not photograph it. Then it was installed in the ship's hidden hold, called the "moon pool," from below by using a submerged barge.

Clementine was positioned over the submarine and then its hydraulic jaws were closed to grab the vessel. Cameras relayed pictures to the ship's crew. Everything had gone according to plan so far. Soviet ships circled the *Explorer* — one even launched a helicopter to investigate more closely and take photographs — but there was nothing to suggest that *Glomar Explorer* was anything but the drillship she was supposed to be. Somehow, the Soviet Union had found out that a secret operation was about to happen, but they didn't know whether this was it. They didn't know precisely where their lost submarine was, either, so they could not tell whether *Glomar Explorer* was parked over it. Even if she was, Soviet military engineers thought it would be impossible to recover the submarine from such a great depth.

Then something went wrong. With the submarine a third of the way up to the surface, *Clementine*'s hydraulic jaws began to fail. An unsupported part of the submarine broke away and fell back to the seabed. Only the front 38 feet (just over 11 m) of the submarine was successfully lifted into the ship's moon pool. It contained the remains of six crewmen, who were buried at sea with full military honors (film of this was later handed over to the Russians). The parts of the submarine the CIA was most interested in — the missile room, control room, engine room and communications equipment — were lost. The Soviet Union believed that two of its nuclear-armed torpedoes had been recovered by the CIA, and this is thought to be consistent with the levels of radiation reportedly measured by *Glomar Explorer*'s crew.

RIGHT: *A lifting frame was lowered from the* Glomar Explorer *to the submarine, which was then raised and taken into a secret compartment in the ship through doors in the hull.*

GLOMAR CHALLENGER

Glomar Explorer is often confused with a deep-sea research and drilling ship called *Glomar Challenger*. Named after the 19th-century oceanographic survey vessel, HMS *Challenger*, *Glomar Challenger* spent 15 years from the late 1960s to the early 1980s drilling almost 20,000 core samples in the Atlantic, Pacific and Indian oceans and the Mediterranean and Red seas. The cores drilled in the Atlantic Ocean provided proof for the theories of seafloor spreading, plate tectonics, and continental drift.

ABOVE: *Often confused with* Glomar Explorer, Glomar Challenger *was a genuine scientific drilling vessel that was not engaged in cloak-and-dagger missions.*

The Secret is Leaked

Project Azorian did not stay secret for long. Just before the mission began, there was a break-in at the offices of a business known as the Summa Corporation. It was part of the Hughes business empire. Some of the documents that were stolen linked the aviator and businessman Howard Hughes to the CIA and *Glomar Explorer*. The CIA was so disturbed by the theft that they involved the FBI, who asked the Los Angeles Police to investigate. Inevitably, the story was eventually leaked to the media. The CIA appealed to the media to keep it quiet. Despite this, the *Los Angeles Times* newspaper ran a story about the mission, which they called Project Jennifer, on February 18, 1975. Television reporters picked up the reports and broadcast the story too. With *Glomar Explorer*'s cover blown, a follow-up mission to recover the rest of the submarine was canceled.

Glomar Explorer was mothballed by the navy for the next 20 years. In the late 1990s, she was converted into a genuine deep-sea drilling ship and leased to Global Marine Drilling. After a series of mergers and takeovers, Global Marine Drilling became Transocean Inc., which bought *Glomar Explorer*. In April 2015 they announced that the ship, now called *GSF Explorer*, was to be scrapped.

MS ALLURE OF THE SEAS

Throughout the 19th and 20th centuries, ocean liners grew bigger and bigger. At the beginning of the 21st century, a new ship broke all previous size records. The *Allure of the Seas* is the biggest passenger ship ever built. It is as long and as heavy as the biggest nuclear-powered aircraft carriers and if it were to stand on end, it would be 124 feet (38 m) taller than the Eiffel Tower.

TYPE: *Oasis*-class cruise ship

LAUNCHED: Turku, Finland, 2009

LENGTH: 1,187 ft (362 m)

TONNAGE: 225,282 gross tons

CONSTRUCTION: welded steel

PROPULSION: three Wärtsilä 18,590 hp (13,860 kW) 12V46D diesel engines plus three Wärtsilä 24,780 hp (18,480 kW) 16V46D diesel engines driving generators to power three Azipods and four bow thrusters

Intercontinental travel by ocean liner went into decline in the 1960s with the advent of jet airliners. But although passengers abandoned point-to-point travel by sea, cruising for pleasure grew in popularity. Most of the old liners were not suitable for cruising. Cruise ships need a shallower draft (hull depth below the waterline) so that they can enter smaller ports and harbors, and they need more cabins with a sea view than a typical liner offers. The design of the new generation of cruise ships placed a greater emphasis on entertainment and amenities. The cruise ship became a vacation experience in itself — a floating resort, instead of merely a means to get from place to place.

RIGHT: Allure of the Seas *is so big that it has its own "Central Park" running down the middle of the ship, with thousands of living trees and other plants.*

Fifty Ships that Changed the Course of History

214

ABOVE: *Giant cruise ships like* Allure of the Seas, *seen here at her home port of Fort Lauderdale, resemble floating apartment blocks. The ship is designed to give the maximum number of cabins a sea view.*

As of 2015, the biggest passenger ships ever built are the *Oasis*-class cruise ships, *Oasis of the Seas* and *Allure of the Seas*. A third *Oasis*-class ship, *Harmony of the Seas*, will join them in 2016 and a fourth is under construction. Their displacement of about 98,420 long tons (100,000 metric tons) is roughly comparable to that of the world's biggest warships. Although the two *Oasis*-class ships already in service were built to the same design, *Allure of the Seas* is 2 inches (50 mm) longer than the *Oasis* and so she holds the current size record. The *Allure* was built in Turku, Finland, in 2008–2009 and launched on November 20, 2009. A year later, she departed for her maiden voyage.

THE PROBLEM OF PIRACY

The *Allure of the Seas* specializes in Caribbean cruises. Nearly 400 years ago, the area was a hotbed of piracy. Pirates like Blackbeard and Captain Kidd preyed on Spanish, French and English ships carrying cargoes of gold and silver. Piracy in the Caribbean was stamped out in the 1830s, but it is still a real risk in other parts of the world. Modern pirates rarely attack cruise ships, preferring freighters with valuable cargoes and small crews. Armed with automatic weapons, they approach in small, fast boats and then board the ship. They hold the crew, ship and cargo for ransom. The Somali coast, Malacca Strait, Gulf of Guinea and South China Sea are high-risk areas. Despite naval patrols and antipiracy measures taken by merchant ships themselves, piracy is likely to be a continuing danger in some places.

Hunkering Down

In order to leave the Baltic and set out across the Atlantic to her home port of Port Everglades, Florida, she had to negotiate Denmark's Great Belt Bridge. Ships pass under a span of the bridge with a clearance of 213 feet (65 m). Unfortunately, the *Allure of the Seas* measures 236 feet (72 m) from the waterline to the top of her smokestacks. The task seemed impossible. Had the ship's builders made a disastrous mistake? As the *Allure* headed for the bridge, her captain gave three commands. First, he called for the ship's retractable smokestacks to be lowered. Second, he ordered the ballast tanks to be flooded with thousands of tons of water to make the ship settle lower in the water. Finally, he accelerated the ship to 20 knots (23 mph or 37 km/h). In shallow water, this produces something called the "squat" effect." Water flowing between the ship and the seabed has to speed up to squeeze through the gap. This lowers the water pressure and sucks the ship downward. As a result, the *Allure of the Seas* passed safely underneath the bridge.

ABOVE: *In 2010, the* Allure of the Seas *passed under Denmark's Great Belt Bridge, even though she was 23 feet (7 m) higher than the bridge!*

FUTURE SHIPS

From time to time, there have been plans for ships significantly bigger than the *Allure of the Seas*, even floating cities where people might live permanently. So far, none have been built. It is difficult to imagine ships much bigger than the *Allure of the Seas* and the other *Oasis*-class ships. Building bigger would deny the ships access to some of the ports a cruise ship might normally visit. Military vessels may have reached their size limit, too. The U.S. *Nimitz*-class and *Gerald R. Ford*-class aircraft carriers are the biggest military vessels afloat today. We are unlikely to see even bigger naval ships, because big ships are also big targets. The invention of the torpedo and guided missile shows that very big ships can be destroyed by very small weapons, so a warship's size is no guarantee of survival. We may already have seen the biggest ships that will ever be built.

Propulsion by Pods

The propulsion system of ships like the *Allure of the Seas* is completely different from that of the old ocean liners. Traditionally, an engine or electric motor turned a shaft, which penetrated the hull and turned a propeller at the end. Steering was achieved by means of a rudder. Modern ships such as *Allure of the Seas* are propelled and steered by units called Azipods. The pods are fitted under the ship's stern. Each pod contains an electric motor driving a propeller. Turning the whole pod steers the ship, so there is no need for a rudder. The *Allure of the Seas* has three Azipods, each with a 20-foot (6-m) propeller.

Unlike a conventional ship's propellers, which push the ship forward, an Azipod's propeller is

RIGHT: *Rudders are out and Azipods are in. The* Allure of the Seas *is propelled and steered by swivelling Azipods.*

BELOW: *The* Allure of the Seas's Aqua Theater *is a leisure pool by day and a theater at night when it hosts aquatic shows with high divers and acrobats.*

usually at the front of the pod, pulling the ship through the water. In the conventional layout, water has to flow around a ship's propeller shaft and its framing before it reaches the propeller. The obstacles in its way disturb the smooth flow of the water. The Azipod propeller, however, has nothing in front of it to disturb the water flow. This improves the propeller's efficiency by 5 percent or more. The *Allure* also has four bow thrusters to help the ship maneuver in the confined space of a small port. Each thruster is 10 times more powerful than a Formula 1 racing car.

The towering vessel has 18 decks — 16 for the ship's 5,492 passengers and two for the crew of 2,384. Facilities for the *Allure*'s passengers include more than a dozen restaurants, four pools, jacuzzis, two FlowRiders (surfing machines), shops, theaters, sports areas and gyms, a tree-lined park and the first zipline to be installed on any cruise ship.

The *Allure of the Seas* is a giant among ships, but she is set to lose her record as the world's biggest passenger ship in 2016 when the third *Oasis*-class ship, *Harmony of the Seas*, is expected to join the fleet. *Harmony* is reported to be longer than the *Allure* by 6 feet (2 m).

FURTHER READING

GENERAL

Alexander, Caroline. *The* Bounty: *The True Story of the Mutiny on the* Bounty. Harper Perennial, 2004.

Ballard, Robert D. T*he Discovery of the* Titanic. Hodder & Stoughton, 1987.

——. *Exploring the* Lusitania: *Probing the Mysteries of the Sinking that Changed History.* Weidenfeld and Nicolson, 1995.

——. *The Lost Wreck of the* Isis. Madison Press, 1990.

Bergreen, Laurence. *Columbus: The Four Voyages 1492–1504.* Penguin, 2013.

Cawthorne, Nigel. *Shipwrecks: Disasters of the Deep Seas.* Arcturus, 2013.

Frame, Chris, and Cross, Rachelle. *The Evolution of the Transatlantic Liner.* History Press, 2013.

Giggal, Kenneth (illus. Cornelis de Vriés). *Great Classic Sailing Ships.* Webb & Bower, 1988.

Griffiths, Denis; Lambert, Andrew; Walker, Fred. *Brunel's Ships.* Chatham Publishing, 1999.

Hough, Richard. *Captain James Cook: A Biography.* Coronet, 2003.

Ireland, Bernard. *The Hamlyn History of Ships.* Hamlyn, 1999.

Jefferson, Sam. *Clipper Ships and the Golden Age of Sail: Races and Rivalries on the Nineteenth Century High Seas.* Adlard Coles, 2014.

Kentley, Eric. Cutty Sark: *The Last of the Tea Clippers.* Conway, 2014.

Lavery, Brian. *The Conquest of the Ocean: The Illustrated History of Seafaring.* Dorling Kindersley, 2013.

——. *Ship: 5,000 Years of Maritime Adventure.* Dorling Kindersley, 2004.

Payne, Lincoln. *The Sea and Civilization: A Maritime History of the World.* Atlantic Books, 2014.

Philbrick, Nathaniel. *In the Heart of the Sea: The Epic True Story That Inspired "Moby Dick."* Harper Perennial, 2005.

Rayner, Ranulf. *The Story of the America's Cup 1851–2013.* Antique Collector's Club, 2015.

Rediker, Marcus. *The* Amistad *Rebellion: An Atlantic Odyssey of Slavery and Freedom.* Verso, 2013.

EXPLORATION

Alexander, Caroline. *The* Endurance: *Shackleton's Legendary Antarctic Expedition.* Bloomsbury, 1999.

Ballard, Robert D. *Adventures in Ocean Exploration: From the Discovery of the "Titanic" to the Search for Noah's Flood.* National Geographic, 2001.

Barrie, David. *Sextant: A Voyage Guided by the Stars and the Men Who Mapped the World's Oceans.* William Collins, 2015.

Fernández-Armesto, Felipe (ed.). *The Times Atlas of World Exploration.* HarperCollins, 1991.

Keay, John (gen. ed.). *The Royal Geographical Society History of World Exploration.* Hamlyn, 1991.

Lincoln, Margarette (ed.). *Science and Exploration in the Pacific: European Voyages to the Southern Oceans in the Eighteenth Century.* Boydell, 2001.

Sobel, Dava. *Longitude: The True Story of a Lone Genius Who Solved the Greatest Scientific Problem of His Time.* Walker, 1995.

SUBMARINES

Bak, Richard. *The CSS* Hunley: *The Greatest Undersea Adventure of the Civil War.* Cooper Square Press, 2003.

Hoyt, Edwin P. *The Voyage of the* Hunley: *The Chronicle of the Pathbreaking Confederate Submarine.* Burford Books, 2002.

Hutchinson, Robert. *Jane's Submarines: War Beneath the Waves from 1776 to the Present Day.* HarperCollins, 2001.

Polmar, Norman, and White, Michael. *Project Azorian: The CIA and the Raising of the K-129.* Naval Institute Press, 2012.

WARSHIPS

Ballard, Robert D. *The Discovery of the* Bismarck. Hodder & Stoughton, 1990.

——. *The Lost Ships of Guadalcanal: Exploring the Ghost Fleet of the South Atlantic.* Weidenfeld & Nicolson, 1993.

Hore, Peter. *Battleships.* Lorenz, 2014.

Ireland, Bernard. *Jane's Battleships of the 20th Century.* HarperCollins, 1996.

——, and Grove, Eric. *Jane's War at Sea.* HarperCollins, 1997.

McGowan, Alan. HMS *"Victory": Her Construction, Career and Restoration.* Caxton, 2003.

Nelson, James L. *Reign of Iron: The Story of the First Ironclads, The* Monitor *and the* Merrimack. Harper Perennial, 2005.

Ross, David. *The World's Greatest Battleships: Illustrated History.* Amber Books, 2013.

Rule, Margaret. *"Mary Rose": The Excavation and Raising of Henry VIII's Flagship.* Conway Maritime Press, 1983.

Walker, Sally M. *Secrets of a Civil War Submarine: Solving the Mysteries of the H. L. Hunley.* Carolrhoda Books, 2005.

USEFUL WEBSITES

America's Navy
www.navy.mil

Battleship *Missouri* Memorial
www.ussmissouri.org

Brunel's SS *Great Britain*
www.ssgreatbritain.org

"*Calypso*." Cousteau: Custodians
of the Sea Since 1943.
www.cousteau.org/who/calypso

Cartwright, Mark. "Trireme."
Ancient History Encyclopedia.
www.ancient.eu/trireme

"Caravels: Blue Water Sailing Ships."
InDepthInfo.com.
www.indepthinfo.com/articles/
caravel.htm

The Great Ocean Liners
www.thegreatoceanliners.com

First Fleet Fellowship Victoria Inc.
www.firstfleetfellowship.org.au

The *Fram* Museum
www.frammuseum.no

Hadingham, Evan. "Ancient
Chinese Explorers." Nova, PBS.
www.pbs.org/wgbh/nova/ancient/
ancient-chinese-explorers.html

"History of *Cutty Sark*."
Royal Museums Greenwich.
www.rmg.co.uk/cutty-sark/history

"HMS *Beagle* Voyage."
AboutDarwin.com.
www.aboutdarwin.com/voyage/
voyage01.html

HMS *Warrrior*
www.hmswarrior.org

"Inventors." About.com.
www.inventors.about.com

Joshua Slocum Society International
www.joshuaslocumsocietyintl.org

The *Kon-Tiki* Museum
www.kon-tiki.no

"Liberty Ships." GlobalSecurity.org.
www.globalsecurity.org/military/
systems/ship/liberty-ships.htm

"*Lusitania*." PBS: Lost Liners.
www..pbs.org/lostliners/
lusitania.html

The Mariners' Museum and Park
www.marinersmuseum.org

The *Mary Rose* Museum
www.maryrose.org

"The *Mayflower*." History.com.
www.history.com/topics/mayflower

McCue, Gary W. "John Philip
Holland (1841–1914) and
his Submarines."
www.reocities.com/pentagon/
barracks/1401

"Military History." About.com
www.militaryhistory.about.com

"Mutiny on the *Bounty*."
New World Encyclopedia.
www.newworldencyclopedia.org/
entry/Mutiny_on_the_Bounty

National Historic Ships UK
www.nationalhistoricships.org.uk

National Maritime Museum Cornwall
www.nmmc.co.uk

National Oceanic and
Atmospheric Administration
www.oceanexplorer.noaa.gov

Naval History Blog
www.navalhistory.org

NavSource Naval History
www.navsource.org

New York Yacht Club
www.nyyc.org

N.S. *Savannah*
www.nssavannah.net

Nydam Mose
www.nydam.nu

"Origins of the *Titanic* Iceberg."
BBC History.
www.bbc.co.uk/history/topics/
iceberg_sank_titanic

"Project 92M *Lenin*." Global Security.
www.globalsecurity.org/military/
world/russia/92m.htm

"*Rainbow Warrior*." Greenpeace.
http://www.greenpeace.org/
international/en/about/ships/
the-rainbow-warrior

"Steaming Across the Atlantic."
ConnecticutHistory.org.
www.connecticuthistory.org/
steaming-across-the-atlantic

Submarine Force Museum
www.ussnautilus.org

"The *Torrey Canyon*'s Last Voyage."
Splash Maritime Training.
www.splashmaritime.com.au/
Marops/data/less/Poll/torreycan.htm

Tri-Coastal Marine
www.tricoastal.com

"*Turbinia*." National Historic
Ships UK.
www.nationalhistoricships.
org.uk/register/138/turbinia

"The Turtle Ship."
www.navy.memorieshop.com/
Korea/index.html

Uboat.net
www.uboat.net

U.S. Carriers
www.uscarriers.net

Woods Hole Oceanographic
Institution
www.whoi.edu

World Shipping Council
www.worldshipping.org

"*Yamato*." World War II Database.
www.ww2db.com/ship_spec.
php?ship_id=B1

INDEX

IMAGE CREDITS

A NOTE ON TONNAGE

Over the centuries, the size and weight of ships has been measured in a bewildering array of different units. Some "tonnage" figures don't refer to weight at all — they are a measurement of its cargo-carrying capacity. They date back to the 14th century, when taxes were first levied on ships according to how much cargo they could carry. The basic unit of cargo was a barrel of wine called a tun. It contained 252 gallons (1,146 l) and weighed 2,240 pounds, or one "long ton" (1.016 metric tons). Until the 19th century, a ship's capacity was often given in tons burden, or burthen. Of many other measurements used for rating ships, the most common include the following:

Displacement tonnage, or displacement, is the weight of water a floating ship displaces. This is the same as the weight of the ship.

Gross register tonnage, introduced in the middle of the 19th century, is a measurement of the total internal volume of a ship. One register ton is a volume of 100 cubic feet (about 2.83 m³). It is calculated by dividing the volume of the ship, measured in cubic feet, by 100.

Gross tonnage has replaced gross register tonnage following an international agreement in 1969. It is given by a mathematical formula based on a ship's volume in cubic meters.